One of the most versatile vehicles produced by Germany during the Second World War period was the 3-ton series halftrack developed by Borgward AG at the company's Hansa-Lloyd Werke which, before the war, had been famous for the manufacture of luxury cars. The new design was referred to as the Hansa-Lloyd kleine 5 (HL.kl 5) and from late 1937 production was taken over by Hannoversche Maschinenbau AG, or Hanomag, but not before 505 vehicles had left the Borgward assembly lines (1). First entering service in 1938 as the Sdkfz 11, the new halftrack provided cross-country transport for a range of medium towed guns including the 10.5cm leFH howitzer and 15cm Nebelwerfer multiple rocket launcher. But it is the armoured version which was based on the chassis of this halftrack, classified as the Sdkfz 251, which is perhaps the best known.

Initially conceived as a means of transporting a squad of infantry to the battlefield while protecting them from small arms fire, the basic design was produced in four main versions and twenty-three variants. By the end of the war over 15,000 examples had been built by Hanomag and also Adlerwerke, Horch Automobilewerke, Skoda, and Borgward (2). The first production vehicles were allocated to 1.Panzer-Divison and took part in the Polish Campaign of 1939, proving their worth from the first days of the fighting.

But plans to completely equip the armoured infantry regiments of Panzer and Panzergrenadier formations with Sdkfz 251 and the lighter Sdkfz 250 halftracks were never realised and most divisions contained only one fully armoured, or Gepanzert, battalion (3). By the end of 1941, when the German army had been fighting in Russia for six months, almost 1,300 Sdkfz 251 halftracks had been produced (4). The Sdkfz 251 ausf C, with its much simplified hull, had been introduced in January of the same year.

As early as the French Campaign of 1940 it had been realised that the army's main battle tank, the Pzkpfw III, was at a decided disadvantage when faced by heavier enemy types such as the British Matilda II and the search began for a replacement. As the turret diameter of the Pzkpfw III prohibited the installation of a larger and more powerful gun, it was decided, just weeks prior to the invasion of the Soviet Union, to experiment with rearming the Pzkpfw IV medium tank which was at that time fitted with the short-barrel 7.5cm L/24 gun. The battles fought on the Eastern Front further demonstrated the inadequacy of the Panzers when matched against the Soviet T-34 and KV series of heavy tanks and by early 1942 production models of the Pzkpfw IV began arriving at the front armed with the long-barrel 7.5cm L/43 gun. This weapon, when employed with the Panzergranate 39 round, was capable of penetrating 77mm of armour plate at ranges of almost 2,000 metres. But these tanks had originally been designed as infantry support weapons and a replacement would have to be found to fill that role. The answer presented itself almost immediately in the shape of the hundreds of 7.5cm L/24 guns which were now no longer required for tank production (5). The Heereswaffenamt, the branch of the army responsible for the development and procurement of weapons, ordered that these guns be fitted to Sdkfz 233 armoured cars and the Sdkfz 251 halftrack.

The firm of Büssing-NAG was contracted to convert existing halftracks at the company's Berlin plant and two prototype vehicles were sent to the Eastern Front in June 1942. The trials were so successful that an order was immediately placed for 150 vehicles and series production of the Sdkfz 251/9 began in October or November of the same year. The first vehicles were allocated to the heavy weapons platoons of armoured infantry companies and these halftracks are first mentioned in Kriegsstarkenachweisungen (KstN) 1125(gp) of 21 December 1942 (6). This was soon extended to the heavy companies of Panzer-Aufklärungs battalions and the heavy companies of Gepanzert infantry battalions. The establishments of these units is explained in greater detail in the section which begins on page 56. Officially referred to as the Kanonenwagen, the halftrack was soon christened Stummel, or stump, by it crews. From September 1943 al conversions were based on the Sdkfz 25 ausf D and in December a new gun moun referred to as the 7.5cm KwK 51(Sf), wa introduced. This featured highe armoured sides and could be fitted to existing vehicles by units in the field and so was also issued in kit form.

At a conference held on 28 October 1944, the Generalinspekteur der Panzertruppen informed Hitler that the Pzkpfw IV with its 7.5cm L/48 gun, the most numerous tank then in service with the German military, was no longer a match for the main Allied types. Experiments in fitting the tank with the 7.5cm L/70 were underway and although this was later found to be unworkable it did raise the possibility that the Heereswaffenamt would be left with a large number of surplus L/48 weapons. Hitler immediately suggested that these be fitted to as many vehicle types as possible, specifically mentioning the Sdkfz 251 halftrack, and going on to say that not only could the L/48 gun be incorporated during production but that it could also be manufactured in kit form to replace the weapons currently in service as Sdkfz 251/9 vehicles. In the following month the Führer again raised the subject with

Notes

1. Somewhat confusingly, the classification as 3-ton refers to the vehicle's capacity or payload while kl 5 indicates the actual weight, which was in fact increased to 6 tonnes by the modifications made at Hanomag. The firms of Büssing-NAG and Deutsche Werke were also involved in the early development of the 3-ton halftracks and I have been unable to discover why production was turned over to Hanomag.

2. From 1938 Skoda, a Czech company, had been incorporated into the immense Reichswerke Hermann Göring. After 1945 Horch Automobilewerke was transformed into Audi. Smaller companies, such as Schöller Bleckmann in Austria, were involved in component manufacture.

3. There were exceptions to this of course but they were unusual or temporary in nature and are mentioned in the unit histories.

4. I should mention that this figure, given by Reichsminister Speer in 1945, has been disputed with a total of just over 900 suggested.

5. The practice of measuring and naming guns [illegible] bore. When used by the German military of the Second World War period the number was arrived at by dividing the total length of the gun, including the breech, by the inside diameter and expressed as L, for example L/24.

6. KstN were detailed organisational charts and tables and their use is explained from page 56.

Notes

1. The plan to end the production life of the Pzkpfw IV is examined in detail in *TankCraft 26, Jagdpanzer IV: German Army & Waffen-SS Tank Destroyers Western Front 1944-1945.*

2. At least one source claims that the Görlitz plant was closed in March 1945 but the Soviet occupation forces found it necessary to prohibit the employees from returning to their jobs after the German surrender on 8 May. The facilities were later taken over by the Russians as a tank repair depot.

Speer, his armaments minister, and it was agreed that as the decision had been taken to end production of the Pzkpfw IV and as the guns could be fitted to Sdkfz 251 halftracks with very little modification, the project should go ahead (1).

Complete vehicles were assembled by Waggon und Maschinenbau AG (Wumag) at the company's Görlitz factory and conversion kits were built by Ardelt-Werke GmbH at Eberswalde. Between December 1944 and February 1945 a total of ninety-five vehicles left the Görlitz plant while perhaps more than 100 conversions sets were completed (2). Figures for March and April are not available but as Eberswalde was not lost until 22 April 1945 and Görlitz was still in German hands as late as 7 May it is possible that small quantities may have been produced in those months.

The first Sdkfz 251/22 halftracks were shipped to units from the Heereszeugamt at some time in December 1944. Of the seventy-two conversion sets that Hitler had demanded be made available immediately, a total of sixty had been sent to units operating on the Western Front as early as 4 December while a further six had been shipped to Cottbus, near the present-day Polish border, and these were earmarked for the Führer-Grenadier-Brigade. Another thirty-two sets were waiting at Eberswalde for shipment and it was expected that an additional thirty-four would be completed by 7 December 1944.

Unfortunately for the modern researcher many formations continued to report the converted vehicles as Sdkfz 251/9 halftracks although they now mounted the larger L/46 gun. In addition, few formations submitted detailed monthly status reports after March 1945 and it should also be mentioned that some conversions were field modifications where towed 7.5cm guns were removed from their wheeled carriage and adapted to fit existing halftracks. Therefore the allocations mentioned in the following unit histories should be treated, in most cases, as provisional only.

KANONENWAGEN AND PAKWAGEN PRODUCTION

Month	Sdkfz 251/9 ausf C	Sdkfz 251/9 ausf D	Sdkfz 251/22	Sdkfz 251/22 Conversion sets	Total Sdkfz 251 Production
March 1943	95 (1)				150
April 1943	30				230
May 1943	31				251
June 1943	0				259
July 1943		24			303
August 1943		6			432 (5)
September 1943		65			627
October 1943		59			526
November 1943		76			657
December 1943		65 (2)			572
January 1944		21			554
February 1944		120			723
March 1944		45			842
April 1944		22			645
May 1944		5			669
June 1944		60			600
July 1944		73			568
August 1944		116			837
September 1944		86			754
October 1944		96			600
November 1944		27		98 (4)	521
December 1944		26	40		466
January 1945		0	41		483
February 1945		3	14 (3)		360
March 1945					173

1. Accurate figures are unfortunately unavailable for individual months before this time. The figure given here is the total production up to the end of March 1943. 2. The later mount is introduced for the 7.5cm L/24. 3. This figure is conjectural and is based on the number of vehicles accepted by the Heereswaffenamt in the previous months as compared to total production to that date. 4. The figure is based on sixty-six sets known to have been shipped by 4 December 1944 and the thirty-two sets reported on hand at Eberswalde in the following week. It is not known if another planned thirty-four sets were completed. 5. Production of the Sdkfz 251 ausf C ends.

Sdkfz 251/9 halftracks were allocated to Panzer divisions and equipped the Schwere Kompanien, or heavy companies, of Panzergrenadier and Panzer-Aufklärungs battalions. The organisation of these units and their place within the division is shown in the accompanying charts. There were of course anomalies and where these are known I have noted them.

Unfortunately for the modern researcher many formations, when reporting the numbers of halftracks on hand, referred simply to these vehicles as Sdfkz or SPW (Sonderpanzerwagen) without mentioning the exact type or model. Similarly, there is sometimes no differentiation made between Sdkfz 251/9 and Sdkfz 251/22 halftracks. In addition, few formations submitted detailed monthly status reports after March 1945 and much of the information we have regarding the allocation of these vehicles is based on sources other than official documentation. Therefore the following lists should be treated, in most cases, as provisional only.

1.Panzer-Division. In early 1944 this formation was fighting in southern Ukraine and during the massive Soviet summer offensives was pushed back as far as the Vistula River. In October the division was sent to Hungary and took part in the defensive battles there, gradually withdrawing into Austria.

On 30 December 1944 the division were on hand and as Panzer-Aufklärungs-Abteilung 1 still fielded towed anti-tank guns in its heavy platoon, these vehicles may have been allocated to I.Bataillon, Panzegrenadier-Regiment 113 which was supposed to be fully armoured.

2.Panzer-Division. The division took part in the battles on the Eastern Front and in early 1944 was sent to France to rest and refit. The division was engaged in the Normandy battles and was almost completely destroyed. Rebuilt in September 1944, 2.Panzer-Division fought in the Ardennes and in the defence of the Rhine, surrendering to units of the US Army at Plauen in May 1945.

In June 1944 the first battalions of Panzergrenadier-Regiment 2 and Panzergrenadier-Regiment 304 were equipped with twelve Sdkfz 251/9 halftracks each, while 5.Kompanie, Panzer-Aufklärungs-Abteilung 2 had six 7.5cm-armed vehicles. On 16 December 1944, on the eve of the Ardennes Offensive, Panzergrenadier-Regiment 2 had eight serviceable 7.5cm-armed halftracks available.

On 30 December 1944 the division reported that seven serviceable Sdkfz 251/9 halftracks were on hand with a further eight in transit. The division does not seem to have been allocated any Sdkfz 251/22 halftracks or any kits to convert the existing Sdkfz 251/9 vehicles. A halftrack of the division is shown on page 23 of the Camouflage & Markings section.

Although this photograph was taken before the period covered by this book it nevertheless provides a good illustration of a Kanonenwagen based on the Sdkfz 251 ausf C in action. Pictured during the winter of 1942-1943, these vehicles are from 8.Kompanie, Panzer-Regiment 6 of 7.Panzer-Division. Notice the tactical sign denoting an armoured infantry company and the number 8 on the hull front. The unusual unit insignia was only used for a short time.

The only definite information that we have about this photograph of an Sdkfz 251/22 halftrack is that it was taken in Czechoslovakia in the spring of 1945. Several units were equipped with these vehicles and fought in the last battles in what is today the Czech Republic and possible candidates include Panzergrenadier-Regimenten 20, 74 and 146 and Panzer-Aufklärungs-Abteilungen 16 and 19. All these formations are mentioned in the main text. This vehicle is also shown in the Camouflage & Markings section on page 25

3.Panzer-Division. The division served on the central sector of the Eastern Front, most notably at Kharkov and along the Dnieper, withdrawing through Ukraine in 1944 into Poland. After taking part in the fighting for the Narew bridgehead near Warsaw, the division was transferred to Hungary in early December. In January 1945 the division was engaged in the unsuccessful attempts to relieve Budapest and Operation Frühlingserwachen, the last major German offensive of the war. On 30 December 1944 the division reported that eighteen Sdkfz 251/9 halftracks were on hand with a further two in transit, suggesting that both I.Bataillon, Panzergrenadier-Regiment 3, the division's armoured infantry battalion, and Panzer-Aufklärungs-Abteilung were both equipped with these vehicles.

4.Panzer-Division. The division was attached to XXXIX.Panzerkorps in mid-1944 and fought in the central Ukraine, retreating into Poland. The division was involved in the defence of Warsaw and the Vistula crossings, most notably at Radzyim and Wolomin where it was able to stop a major Soviet armoured thrust. Later in 1944 the division was transferred to Lithuania and subordinated to 3.Armee.

Throughout January 1945 the division took part in the defensive battles in West Prussia and was eventually cut off in the Frische Nehrung on the Baltic coast where the survivors surrendered to the Soviet army. The division's armoured infantry battalion, I.Bataillon, Panzergrenadier-Regiment 12, was equipped with Sdkfz 251/9 halftracks and the eleven vehicles on hand in late December were probably all with this unit. A Grundgliederung, or organisational chart, of 1 March 1945 shows that 5.Kompanie, Panzer-Aufklärungs-Abteilung 4 was authorised six Sdkfz 251/9 halftracks but I cannot confirm that these were actually present. Although the records are unclear, the division may have received a number of Sdkfz 251/22 halftracks in February 1945.

5.Panzer-Division. During 1944 the division took part in the battles in Ukraine and central Poland, supporting 5.SS-Panzer-Division Wiking, and was later transferred to Lithuania and then Kurland in August. In November the division was fighting in East Prussia as part of 4.Armee and remained there until the end of the war. The division reported on 30 December 1944 that a total of twenty-five Sdkfz 251/9 halftracks were on hand and it will be noticed that this is far more that the authorised allocation.

During the winter of 1944-1945 the division's Panzergrenadier-Regiment 14 was reinforced and its 4.Kompanie contained thirteen Sdkfz 251/9 halftracks. The first, second and third companies were each equipped with two Sdkfz 251/9 vehicles as normal giving a total of nineteen. I have been unable to determine if a special KstN was issued to cover this organisation but it was probably a temporary solution. It cannot be assumed that the remainder of the 7.5cm-armed halftracks were issued to Panzer-Aufklärungs-Abteilung 5 as photographic evidence confirms that Panzer-Regiment 31 had at least one of these vehicles.

A halftrack of the Panzer regiment is shown on page 24 of the Camouflage & Markings section. It is known that the heavy company of the Aufklärungs battalion was equipped with towed guns well into 1945 and may not have received any 7.5cm-armed halftracks.

6.Panzer-Division. The division took part in the attempts to relieve the Cherkassy Pocket during the winter of 1944-1945 and later fought in the battles around Kamenets-Podolsky.

The division was withdrawn to Germany but returned to the East in July 1944. During the summer and autumn of that year 6.Panzer-Division was engaged in the defensive battles in Poland and East Prussia and in November was transferred to Hungary. The division took part in the battles around Budapest and along the Garam River where the Sdkfz 251/9 halftracks of 4.Kompanie, Panzer-Aufklärungs-Abteilung 6 were able to stop a Soviet night attack at Nagykalna on 21 December 1944, destroying eight Russian tanks.

On 30 December 1944 the division reported that nineteen Sdkfz 251/9 halftracks were on hand and these were almost certainly with II.Bataillon, Panzergrenadier-Regiment 114 and Panzer-Aufklärungs-Abteilung 6. In early 1945 the division was involved in the defence of Vienna and subsequently withdrew into Czechoslovakia in May 1945. A halftrack vehicle of this formation is shown on page 22 of the Camouflage & Markings section.

7.Panzer-Division. In July 1944 the division, which had been fighting in the Ukraine, was transferred to Lithuania and took part in the series of counterattacks around Memel on the Baltic coast as part of 3.Panzerarmee. The Germans reached the city but were soon after encircled and in November the division was evacuated by sea to Germany, leaving behind all its heavy equipment. On 30 December 1944 the division reported that twenty Sdkfz 251/9 halftracks were on hand which meant a full complement for II.Bataillon, Panzergrenadier-Regiment 6 and one of the companies of Panzer-Aufklärungs-Abteilung 7.

In January 1945 the division had a number of Sdkfz 251/22 halftracks on hand which may have been converted Sdkfz 251/9 vehicles. In the same month 7.Panzer-Division returned to the Eastern Front and fought in northern Poland before being pushed back into Pomerania. In March the division was fighting near Danzig and in April 1945 was again evacuated by sea but this time only a small remnant of the division managed to escape.

On 14 April 1945 an order was issued by the Generalinspekteur der Panzertruppen directing that the division be reorganised according to the Kampfgruppe Panzer Division Grundgliederung issued on 25 March 1945, one of very few formations to be rebuilt using this structure. A vehicle of the division is shown on page 27 of the Camouflage & Markings section.

8.Panzer-Division. During the summer of 1944 the division served on the Eastern Front in the encirclement battles in the Ukraine and eastern Poland, even though it was in the process of rebuilding when the massive Soviet offensive began in June. Withdrawing towards Lviv, the division took part in the battles for the Dukla Pass, at the foot of the Carpathian Mountains, and from December 1944 was involved in the defence of Budapest. At the end of the same month the division reported that sixteen Sdkfz 251/9 halftracks were on hand and as the heavy support platoon of Panzer-Aufklärungs-Abteilung 8 was equipped with towed anti tank guns well into 1945, these may have all been with I. Bataillon, Panzergrenadier-Regiment 8.

In early 1945 the division was moved to Silesia and took part in the unsuccessful relief of Breslau and the recapture of Lauban and was then moved to Czechoslovakia where it remained until the end of the war. There is some evidence that a number of Sdkfz 251/22 halftracks or conversion kits were received in March 1945. A halftrack of this formation is shown on page 28 of the Camouflage & Markings section.

9.Panzer-Division. After fighting on the Eastern Front the division was rebuilt in France and took part in the later stages of the Normandy campaign where it was reduced to one weak battalion of tanks and less than a regiment of infantry.

Reinforced, the division was engaged in the fighting for Geilenkirchen and Aachen with the units of 15.Panzergrenadier-Division. In September 1944 the division's Panzer-Aufklärungs-Abteilung 9 was equipped with five Sdkfz 251/9 halftracks and although these vehicles are not mentioned in a November report, a single halftrack is reported as being on hand in December. The battalion's organisation is shown in the chart on page 6.

On 16 December 1944, the day the Ardennes Offensive commenced, the division's Panzergrenadier-Regiment 10 had only one Sdkfz 251/9 halftrack on hand while Panzergrenadier-Regiment 11

...........*text continued on page 8*

PANZER-AUFKLÄRUNGS-ABTEILUNG 9, SEPTEMBER 1944

Badly depleted, the remnants of 9.Panzer-Division were withdrawn from the Eastern Front in early 1944 and sent to France to be rebuilt and the division's reconnaissance battalion, Panzer-Aufklärungs-Abteilung 9, was reformed during the spring and summer of the same year. By July the battalion was at almost full strength but the later fighting in Normandy and the subsequent retreat took its toll and by September, when the front had stabilised somewhat, the battalion could only muster the assets shown below. As the division did not complete its training until early August 1944 it is likely that the Kriegsstärkenachweisungen (KStN), or organisational charts, issued in April were used but this has not been confirmed and it was not uncommon for older orders to be followed. As a matter of convenience I have anglicised many of the German terms– I should mention that these are approximations only. The KstN charts relating to units equipped with the Sdkfz 251/9 and 251/22 are described in greater detail on pages 56 to 59.

Battalion Headquarters and Headquarters Company — **Maintenance Company**

Headquarters Company Commander | **1st (Signals) Platoon** | **2nd Platoon** | **3rd Platoon**

3 x Sdkfz 251/3, 3 x Sdkfz 250/3

By September 1944 the company reported that only five serviceable Sdkfz 234/1 armoured cars were on hand. Normally one of these vehicles would have been allocated to the company commander and one to each to the second and third platoons which were also authorised one Sdkfz 234/3 armed with the 7.5cm L/24 gun. The latter are recorded as being issued in July 1944 and it is possible that the September report confused the two types. Also in July it seems that the battalion was allocated a total of thirteen Sdkfz 234/1 armoured cars but I can offer no further information as to their fate.

1st Company

Company Commander | **1st Platoon** | **2nd Platoon** | **3rd Platoon** | **4th Platoon**

This battalion and Panzer-Aufklärungs-Abteilung 4 were the only units authorised the fully-tracked version of this organisation which was based on Panzer-Spähkompanie ausf b, KstN 1162b von 1.3.1944 and equipped with the Pzkpfw II ausf L Luchs tanks. The company was made up of four platoons, each allocated six tanks, with a further tank going to the company commander. The battalion actually received twenty-six tanks in July 1944 but by September only eighteen were on hand.

2nd Company

Company Commander | **1st Platoon** | **1st Platoon** | **3rd Platoon** | **4th Platoon**

2 x Sdkfz 250/3 | *2 x Sdkfz 250/10, 6 x Sdkfz 250/1* | *2 x Sdkfz 250/10, 5 x Sdkfz 250/1* | *3 x Sdkfz 250/1*

Officially the battalion's 2.Kompanie would have been formed using KstN 1113 (gp)(fG) von 1.4.1944 but it seems to have built with what was available at the time and may have used older tables of organisation. The Sdkfz 250/10 was armed with the 3.7cm gun and had largely been replaced by the Sdkfz 250/8 by this time. The fourth platoon would normally have been made up of a Granatenwerfer Gruppe equipped with four Sdkfz 250/7 halftracks carry the 8cm Gr.W.34 mortar and a Kanonen Gruppe of two Sdkfz 250/8 and one Sdkfz 250/1 halftracks. In actuality the company's heavy weapons were concentrated in the third platoon which, by September, contained two heavy mortars towed by medium cars and three Sdkfz 250/1 halftracks which moved the platoon's 7.5cm Pak 40 anti-tank guns. It is possible that a fourth platoon was never formed.

3rd Company

Company Commander | **1st Platoon** | **2nd Platoon** | **3rd Platoon** | **4th Platoon**

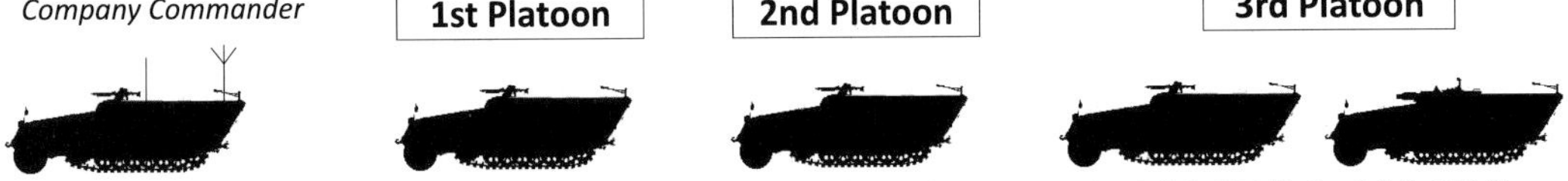

3 x Sdkfz 251/3 | *6 X Sdkfz 251/1* | *6 x Sdkfz 251/1* | *2 x Sdkfz 251/1, 2 x Sdkfz 251/9*

The third company of a reconnaissance battalion was based on KstN 1114 (gp)(fG) von 1.4.1944 which allowed for a headquarters element with two Sdkfz 251/3 communications halftracks and four platoons. The first, second and third platoons were authorised nine Sdkfz 251/1 halftracks and a single Sdkfz 251/17 as a platoon commander's vehicle. This presents something of a mystery as the early Sdkfz 251/17 vehicles, armed with the 2cm Flak 38, were only issued to the Hermann Göring Division and the later Sdkfz 251/17 Schwebelafette halftracks had not entered production in July 1944. In any case the company had none. The fourth platoon provided the heavy support and contained a Granatenwerfer Gruppe with two Sdkfz 251/2 mobile 8cm mortars and a Kanonen Gruppe with two Sdkfz 251/9 halftracks armed with the 7.5cm gun. I cannot confirm that the company was ever fully formed an it appears that by September at least, the third and fourth platoons were amalgamated and only the vehicles shown above were on hand.

4th Company

Company Commander | **1st Platoon** | **2nd Platoon** | **3rd Platoon** | **4th Platoon**

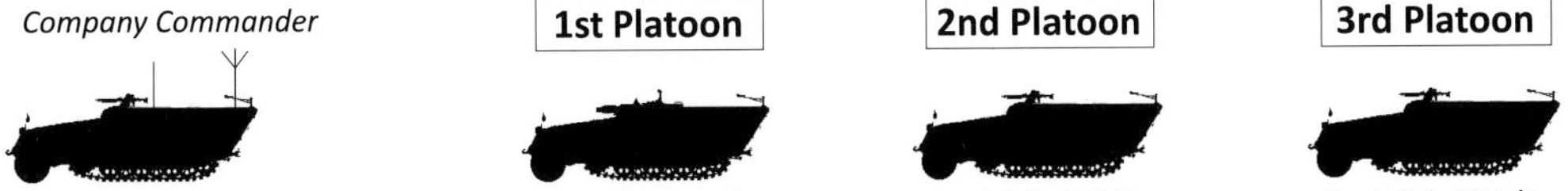

1 x Sdkfz 251/3 | *3 x Sdkfz 251/9* | *3 x Sdkfz 251/7* | *4 x Sdkfz 251/1*

Officially the fourth company was one of the strongest elements of the battalion containing a Pionier-Zug, with mobile bridging equipment, a Granatwerfer-Zug and a Kanonen-Zug. In reality it was made up of a headquarters section with a single Sdkfz 251/3 communications halftrack and three platoons equipped as shown above. None of the authorised 8cm mortars were on hand and the Sdkfz 251/9 halftracks of the fourth platoon were replaced by three 7.5cm Pak 40 towed anti-tank guns.

............text continued from page 5

reported that four were available. By the end of the month the division had a total of fourteen 7.5cm-armed halftracks on hand. In January 1945 the division received a number of Sdkfz 251/22 halftracks and was at that time fighting in the Eifel Mountain area and later at the Remagen Bridgehead.

During the first week of April 1945, the division was reorganised according to the Kampfgruppe Panzer Division Grundgliederung issued on 25 March 1945, one of only four formations known to have been rebuilt using this structure. Most of 9.Panzer-Division was trapped in the Ruhr Pocket in April 1945 but a small Kampfgruppe continued fighting with 11.Armee in the Harz Mountains.

11.Panzer-Division. After it was almost completely destroyed in the Cherkassy Pocket battles on the Eastern Front in early 1944 the division was rebuilt in southern France.

With the Allied landings in August the division began a fighting withdrawal along the Rhône valley to Besançon and then into Alsace. On 1 September 1944, I. Bataillon, Panzergrenadier-Regiment 110 was equipped with a total of five Sdkfz 251/9 halftracks, which was just one short of its full authorisation. At the same time 5.Kompanie, Panzer-Aufklärungs-Abteilung 11 reported that four Sdkfz 251/9 halftracks were available. The division was engaged in the defence of the Belfort Gap and the battle of Arracourt in

Photographed just after the end of the war, this Sdkfz 251/22 halftrack of 11.Panzer-Division is based on the later production version of the basic vehicle as shown by the engine access hatches. The division's famous unit insignia is just visible behind the headlight and it appears that a spare wheel was carried on the hull front at some time. Note the two small repairs on the hull side. This halftrack is also shown in the Camouflage & Markings section on page 25.

A rear view of the halftrack shown on the previous page. The unit insignia, painted onto the hull to the right of the rear access doors, is clearly visible in this image. The traces of a factory-applied camouflage scheme can still be discerned on the hull sides.

late September where the German units suffered particularly heavy casualties. By 16 December 1944, when the division took part in the Ardennes Offensive, the total number of Sdkfz 251/9 halftracks with Panzergrenadier-Regiment 110 had risen to eleven, suggesting that both battalions may have been at least partially armoured.

By the end of the same month, two Sdkfz 251/22 halftracks were in transit to the division and during January 1945 further shipments of vehicles and conversion sets were made (1). but I have been unable to confirm the numbers or how they were allocated within the division.

The division was involved in the attacks on the Remagen Bridgehead in March 1945 but was soon after transferred to southern Germany and subsequently encircled in the Ruhr Pocket. An Sdkfz 251/22 halftrack of this division is shown on page 25 of the Camouflage & Markings section.

12.Panzer-Division. Formed in 1940, the division fought in the East from the invasion of the Soviet Union until the end of the war, surrendering in what is today western Latvia.

In January 1945 the armoured and motorised elements of the division were amalgamated with the remnants of 14.Panzer-Division to form what was initially referred to as Panzer-Aufklärungs-Gruppe Kurland and later Panzer-Brigade Kurland. At that time the division's reconnaissance unit, Panzer-Aufklärungs-Abteilung 12, reported that forty-two Sdkfz 251 halftracks of different types were on hand (1). On 29 January 1945 the division reported that six of its ten Sdkfz 251/9 halftracks had been upgunned with the Pak 40 but the reports are unclear and this may have been accomplished by using the towed anti-tank guns of 3.Kompanie, Panzerjäger-Abteilung 2. This left the anti-tank company with a single 7.5cm Pak 40 gun and the divisional commander requested that the Panzerjäger battalion be allocated a number of assault guns or Jadgpanzers.

The halftracks are not mentioned in the strength report for Panzer-Brigade Kurland of 15 March 1945.

13.Panzer-Division. In early September 1944, after being badly depleted by the battles on the Eastern Front, the division was transferred to Truppenübungsplatz Örkeny south of Budapest to rest and refit and although it absorbed the remnants of Panzer-Brigade 110 it was never rebuilt to its full authorised strength.

Over 100 of the division's armoured halftracks came from Panzer-Brigade 110 and this may have been the origin of the seven Sdkfz 251/9 vehicles reported as being on hand on 30 December 1944. It is known that the division's Panzer-Aufklärungs-Abteilung 13 was equipped with towed anti-tank guns well into 1945 and it would seem likely that the 7.5cm armed halftracks were all allocated to I. Bataillon, Panzergrenadier-Regiment 66. Indeed, photographs taken in Budapest on 11 February 1945 after the division

Notes

1. The battalion is referred to as Gepanzert, or armoured, in the report.

attempted to break out of the city clearly show a number of Sdkfz 251/9 halftracks carrying the regiment's distinctive markings. The survivors of the division were used to build Panzer Division Feldherrnhalle 2 (see below).

14.Panzer-Division. Although this formation had been severely weakened by the fighting in Romania in mid-1944 it was ordered to move over 1,300 kilometres to Schaulen, present-day Siauliai in Lithuania, to take part in the defence of the Baltic States. The division fought in the battles around Riga and in Kurland where it remained until the end of the war.

At the end of December 1944 the division reported that two Sdkfz 251/9 halftracks were on hand and these were probably with I.Bataillon, Panzergrenadier-Regiment 103.

16.Panzer-Division. In July 1944 the division was encircled by Red Army units between Pavlivka and Rykovychi, south of Kovel in northern Ukraine, and although most of 16.Panzer-Division managed to escape, the Germans were forced to fight their way westwards as far as Przemyl in south-eastern Poland. Soon after, the division was thrown into the battles for the Baranow Bridgehead and in early October was withdrawn from the front.

The division reported that sixteen Sdkfz 251/9 halftracks were on hand on 30 December 1944 and these were probably allocated to both Panzer-Aufklärungs-Abteilung 16 and II.Bataillon, Panzergrenadier-Regiment 64, the division's armoured infantry battalion. In January 1945 the division was engaged in the battles for the Steinau Bridgehead on the Oder River and took part in the recapture of Lauban.

In March 1945 the division absorbed the units of Panzer-Division Jüterbog including a number of Sdkfz 251/22 halftracks. With most of its armour gone and with its infantry element reinforced by stragglers and boys of the Hitler Youth, the division nevertheless fought almost to the last day of the war taking part in counterattacks at Rogau in Upper Silesia, today Rogów Opolski, Brno and Opava in Czechoslovakia and Grätz, modern-day Hradec nad Moravicí. A few survivors managed to make their way to the West

A vehicle of this division is shown on page 17 of the Camouflage & Markings section.

17.Panzer-Division. Formed in 1940, the division served on the Eastern Front. The 30 December 1944 report states that nine Sdkfz 251/9 halftracks were on hand at that time but as neither of the division's Panzergrenadier regiments had an armoured infantry component these vehicles were probably allocated to the Panzer-Aufklärungs battalion. In any case most of the heavy equipment was lost in the fighting along the Vistula. The division was rebuilt in January 1945.

At about this time, or perhaps in the following month, twelve Sdkfz 251/22 halftracks were received and these were all concentrated in Panzerjäger-Abteilung 27.

One of the battalion's halftracks is depicted on page 26 of the Camouflage & Markings section.

Below and at right: An Sdkfz 251/9 halftrack of 13.Panzer-Division photographed after the fall of Budapest in February 1945 where much of the division was destroyed. The composite marking shown here, which combines the formation insignia and title with the symbol identifying the type of unit, were increasingly common from late 1944. This example is somewhat conjectural but is based on a clear photograph of an Sdkfz 138/1 Grille ausf M self-propelled gun of the same division and includes the 13.Panzer-Division unit insignia, a tactical symbol identifying the heavy platoon of a Gepanzert infantry company and the number of the company and regiment.

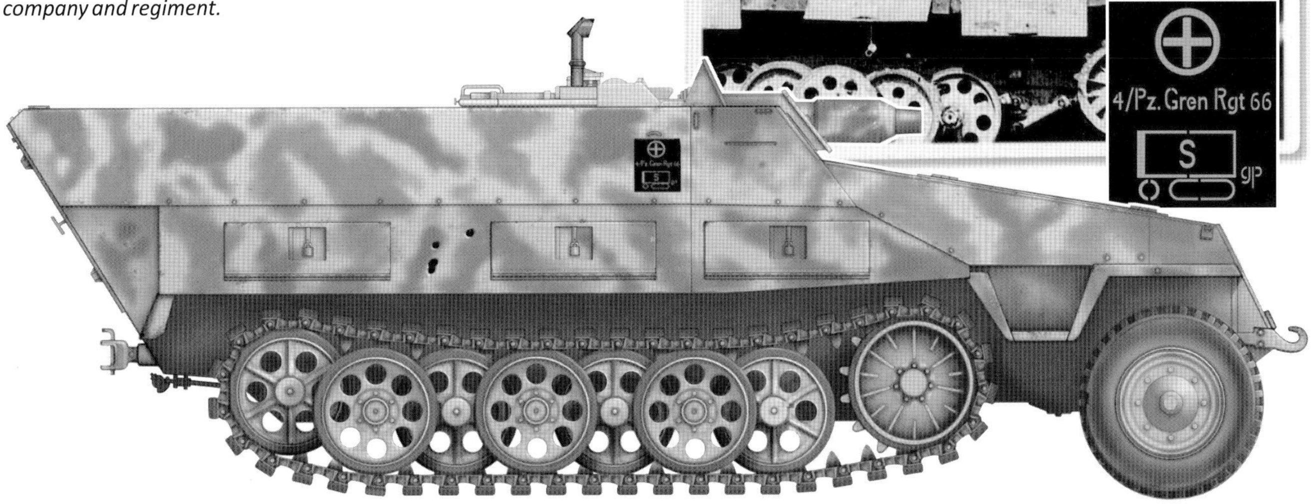

Disabled during the battles in Czechoslovakia in the spring of 1945, this Sdkfz 251/22 is one of twelve issued to Panzerjäger-Abteilung 27 of 17.Panzer-Division. Inset is an enlarged view of the marking on the hull front with the tactical symbol of an anti-tank gun company, enclosing the company number, and the title of the battalion. This vehicle is also depicted in the Camouflage & Markings section on page 26.

19.Panzer-Division. Formed in 1940, the division served exclusively on the Eastern Front. From the summer of 1944 to early 1945, 19.Panzer-Division fought in Poland and then supported the divisions of XXIV.Panzerkorps in Czechoslovakia where the last elements, scattered between Ostrava and Brno, surrendered in May 1945.

On 30 December 1944 the division reported that fifteen Sdkfz 251/9 halftracks were on hand and although I have been unable to definitely determine which units they were allocated to a number were certainly with the Aufklärungs battalion. The division was allocated a total of nineteen Sdkfz 251/22 halftracks in early 1945, probably January, the largest number shipped to any formation. Four of these vehicles served with 3.Kompanie, Panzer-Pionier-Battalion 19 with another four allocated to 4.Kompanie, Panzer-Aufklärungs-Abteilung 19 and eleven going to 4.Kompanie, Panzergrenadier-Regiment 74.

20.Panzer-Division. The division was badly depleted by the battles of summer 1944 on the Eastern Front and from August was first transferred to Romania and then to Truppenübungsplatz Arys in East Prussia to be rebuilt.

The division reported that a total of eighteen Sdkfz 251/9 halftracks were on hand on 30 December and these were almost certainly allocated to Panzer-Aufklärungs-Abteilung 20 and II.Bataillon, Panzergrenadier-Regiment 59. The division was returned to the front and in early 1945 took part in the defensive battles in Upper Silesia, today a province of south-western Poland, notably at Breslau and Schweinitz. In April 1945 the division was involved in a three-day counterattack near Görlitz, which eventually proved unsuccessful, and a further action near Bautzen that relieved the German garrison after inflicting heavy losses on the Russians.

By early May 1945 the remnants of the division had retreated into Czechoslovak territory north-east of Prague and elements were taken into Soviet captivity there.

Most of Panzer-Aufklärungs-Abteilung 20 managed to reach the Allied lines and surrendered to units of the US Army.

21.Panzer-Division. The division was rebuilt in France in late 1943 after it was completely destroyed in North Africa. From its formation the division's Panzer-Aufklärungs-Abteilung was equipped with Sdkfz 251/9 halftracks, reporting that six were on hand with the battalion's 5.Kompanie.

The division was engaged in the Normandy battles where it was so badly depleted that the surviving elements were merged with 16.Luftwaffe Felddivision. In October 1944 the division absorbed the remnants of Panzer-Brigade 112 but it is unlikely that many of the brigade's vehicles survived the battles of September. In the same month Panzer-Aufklärungs-Abteilung 21 reported that just a single 7.5cm-armed halftrack was available.

Although only partly rebuilt the division took part in the Ardennes Offensive. On 30 December the division reported that four

Sdkfz 251/9 halftracks were on hand while a further two were in transit. In early 1945 the division was employed in the drive on Strasbourg as part of Operation Nordwind and in February was transferred to the Eastern Front.

In March 1945 the division was attached to Panzerkorps Grossdeutschland and there is some evidence that a number of Sdkfz 251/9 halftracks were converted to mount the 7.5cm L/46 gun at this time.

23.Panzer-Division. This formation served on the Eastern Front and in mid-1944 took part in the battles for the Sandomierz Bridgehead on the Vistula River in southern Poland as part of 4.Panzerarmee. In September the division was transferred to Hungary and in the following month was engaged in a major counterattack which succeeded in recapturing the town of Nyiregyhaza, claiming the destruction of 600 Russian tanks in the process.

The division remained in Hungary, reporting on 30 December 1944 that ten Sdkfz 251/9 were on hand and these were all allocated to Panzer-Aufklärungs-Abteilung 23. The majority of the division was trapped and destroyed in the area around Stuhlweissenburg, modern-day Székesfehérvár in central Hungary, in March 1945 but some elements, including parts of Panzer-Aufklärungs-Abteilung 23, managed to make their way to Austria where they surrendered to British units.

A halftrack of this division is shown on page 27 of the Camouflage & Markings section.

24.Panzer-Division. This formation was completely destroyed at Stalingrad and rebuilt in mid-1943. The first battalion of the division's Panzergrenadier-Regiment 26 was organised as Gepanzert, or fully armoured, but I have been unable to confirm the allocation of any 7.5cm-armed halftracks. On 30 December 1944 the division reported that five Sdkfz 251/9 halftracks were on hand, probably with Panzer-Aufklärungs-Abteilung 24.

It is possible, although far from certain, that a number of Sdkfz 251/22 halftracks were with the division from January 1945.

25.Panzer-Division. The remnants of the badly-depleted division were withdrawn from the Eastern Front in early 1944 to be rebuilt but in September, before the process was completed, the division was returned to the East and took part in the battles around Warsaw.

In October the division absorbed the surviving elements of Panzer-Brigade 104 but it is unlikely that many of the brigade's vehicles were operational. The division reported on 30 December that fourteen Sdkfz 251/9 halftracks were on hand but it is unclear how these were distributed.

The division's Panzer-Aufklärungs-Abteilung 25 may have had as many as six but this division was unusual in having two armoured infantry battalions (1) and both probably had a number of Sdkfz 251/9 halftracks on hand at some time. The division definitely operated a number of Sdkfz 251/22 halftracks in March 1945 and these may have been received as early as January.

Notes

1. Both I.Bataillon, Panzergrenadier-Regiment 146 and I.Bataillon, Panzergrenadier-Regiment 147 were organised as Gepanzert, or armoured. The latter was disbanded in February 1945.

At left: An Sdkfz 251/9 ausf C of Panzer-Aufklärungs-Abteilung Grossdeutschland probably photographed during the fighting around Wilkowischen, modern-day Wylkowyszki, during August 1944. Note the division's famous Stahlhelm unit insignia, partially visible on the hull front. Above: A later production halftrack of Panzerfüsilier-Regiment Grossdeutschland disabled in spring 1945 when the division was fighting in East Prussia. This vehicle is shown in the Camouflage & Markings section on page 21.

116.Panzer-Division. Formed in early 1944, this formation took part in the battles in Normandy and later the Ardennes Offensive. On 16 December 1944, the first day of the assault, the division reported that nine Sdkfz 251/9 halftracks were on hand with I.Bataillon, Panzergrenadier-Regiment 60. By the end of the month this number had increased to ten and a further three were in transit.

The division's Panzer-Aufklärungs-Abteilung 116 is known to have been equipped with towed anti-tank guns in 1945 and it seems that it was not allocated any 7.5cm-armed halftracks.

It is possible that a number of the halftracks were converted to Sdkfz 251/22 vehicles in February or March 1945 but this has not been confirmed. On 14 April 1945 the division was reorganised according to the Kampfgruppe Panzer Division Grundgliederung issued on 25 March 1945, one of very few formations to be rebuilt using this structure.

Panzer-Lehr-Division. In early 1944 this formation was the strongest armoured division of the German army and was stationed in northern France in expectation of the Allied invasion.

The battalions of Panzergrenadier-Regiment 901 and Panzergrenadier-Regiment 902 were all carried in armoured halftracks and each battalion had six Sdkfz 251/9 vehicles giving a total of twenty-four. Also at this time Panzeraufklärungs-Lehr-Abteilung 130 had six 7.5cm-armed halftracks available with its fifth company.

The division suffered heavily in the Normandy battles and the subsequent retreat and by 3 September 1944, Panzergrenadier-Regiment 901 had just four Sdkfz 251/9 vehicles available while Panzergrenadier-Regiment 902 had only ten armoured halftracks of all type. The reconnaissance battalion seems to have lost all its 7.5cm-armed halftracks by that time. The division was partly rebuilt and returned to the front in November to take part in an abortive counterattack towards the Saverne Gap in the northern Vosges region.

On 16 December 1944, the day the Ardennes Offensive commenced, Panzergrenadier-Regiment 901 reported that seventeen Sdkfz 251/9 were on hand and all these vehicles were still operational at the end of the month. This is the only figure I have been able to confirm for this time and it may be that all the 7.5cm-armed halftracks were concentrated in one regiment. In March 1945 the division received six Sdkfz 251/22 halftracks.

A halftrack of this division is shown on page 19 of the Camouflage & Markings section.

Panzer-Division Feldherrnhalle 1. Formed in late 1943 as a Panzergrenadier division, this formation was sent to the East in January 1944 and was almost completely destroyed in the fighting for Minsk in the following summer.

From September 1944 the division was rebuilt at Truppenübungsplatz Warthelager near modern-day Poznan in Poland and on November 27 1944 was renamed Panzer Division Feldherrnhalle (FHH). During its formation the division absorbed the remnants of Panzer-Brigade 109 but this may have constituted as little as a battalion of infantry. On 30 December 1944 the division reported that seven Sdkfz 251/9 halftracks were on hand with I.Abteilung, Panzergrenadier-Regiment FHH, one more than the battalion's authorised allocation. It appears that the vehicles which would normally have been allocated to the division's Aufklärungs-Abteilung FHH were replaced by Sdkfz 251/21 Drilling halftracks.

The division was sent to Hungary and took part in the defence of Budapest but by Christmas 1944 was cut off in the capital with the rest of IX.Waffen-Gebirgskorps. On 12 February 1945 the encircled Axis units attempted to break out of the city but very few managed to escape and Panzer-Division FHH ceased to exist. The division was once again rebuilt and named Panzer-Division Feldherrnhalle 1 on 10 March 1945 to distinguish it from the reformed 13.Panzer-Division which became Panzer-Division FHH 2 at the same time (see below).

The division's Panzerjäger-Abteilung FHH 1 was well below strength in March 1945 reporting that just six Jagdpanzer 38 were on operational. The same report mentions seven 'Stu.Kan' and these may be some of the nine Sdkfz 251/22 halftracks that were authorised for the battalion's 3.Kompanie. In April the division was surround in a huge pocket in the Deutsch-Brod area east of Prague and completely destroyed.

Panzer-Division Feldherrnhalle 2. This was formed in March 1945 from the survivors of 13.Panzer-Divison, the recently-raised Panzer-Brigade 110, various replacement units and parts of I.Kavallerie-korps .

The division's Panzer-Regiment FHH 2 was authorized three fully armoured Panzergrenadier companies and a Kanonen-zug equipped with six Sdkfz 251/22 halftracks while Panzerjäger-Abteilung FHH 2 was to have a further nine. There is no firm evidence that these halftracks were ever received and although the division's strength reports mention 7.5cm Pak guns they do not specify if they were towed or self-propelled. The division fought in the battles along the upper Danube in April 1945.

It is interesting to note that although both the Feldherrnhalle divisions were raised in early March, they were organised according to the orders of 25 March 1945 for the formation of Panzer or Kampfgruppe Panzer divisions, that is, with a mixed Panzer regiment.

Panzer-Division Jüterbog. Formed on 20 February 1945, the division was little more that a weak brigade. On 26 February the division received six Sdkfz 251/22 halftracks and these were probably allocated to Panzer-Aufklärungs-Kompanie Jüterbog which reported six Schützenpanzerwagen (SPW), or armoured infantry vehicles, described as Schwere or heavy, as being on hand at the time.

On 5 March all elements of the division were absorbed by 16.Panzer-Division.

Panzer-Division Schlesien. Formed as Panzer-Division Döberitz on 20 February 1945, the division, which only reached the strength of a reinforced regiment, was renamed within two days. Late in the same month six Sdkfz 251/22 halftracks were received and the first, second and third companies of I.Bataillon, Panzer-Grenadier-Regiment Schlesien were each allocated two vehicles. The division took part in the fighting around Frankfurt an der Oder but as early as 18 March 1945 it had been decided to incorporate the elements that had been raised so far into Panzer-Division Holstein and this was accomplished by the end of the month.

Almost immediately the survivors of both formations were absorbed by 18.Panzer-Division which was largely destroyed in the fighting on the Oder Front in April 1945.

Panzer-Division Clausewitz. Raised in early April 1945, this was the last Panzer division formed during the war. The division may have had a number of Sdkfz 251/22 halftracks on hand taken over from the partly-formed Panzer-Division Holstein.

The organisation of Panzer-Division Clausewitz is examined in some detail in *TankCraft 18, Panther Tanks: German Army and Waffen-SS Defence of the West, 1945.*

Führer-Begleit-Division. See Führer-Begleit-Brigade below.

Führer-Grenadier-Division. See Führer-Grenadier-Brigade below.

Soldiers of General Walther Wenck' s 12.Armee and civilians wait to cross the River Elbe during the first week of May 1945. On 12 April the retreating Germans had destroyed the Elbebrücke at Tangermünde as US Army units approached from the west and part of the eastern end of the bridge, which runs into the road to Fischbeck, is visible in the background. Although it is not possible to determine the unit to which this Sdkfz 251/22 was allocated, most sources agree that the only armoured formation with 12.Armee at this time was Panzer-Division Clausewitz. However, there is no surviving record of any of these halftracks being on hand with the division and at least one source states that elements of 18.Panzergrenadier-Division, which took over the Sdkfz 251/22 halftracks of Panzer-Division Schlesien, were with 12.Armee as part of Kampfgruppe Reimann. I have been unable to confirm this and it is also possible, but less likely, that some stragglers from 17.Panzer-Division and 21. Panzer-Division were present.

These halftracks of Panzer-Brigade 106 were captured intact after the fighting around Mairy, south-east of Sedan, in early September 1944. It is not known if the brigade's unit insignia, shown at left, was carried by the Sdkfz 251/9 vehicles but it certainly was applied to the Drilling MG 151 halftracks as can be seen in the smaller image. The inscription just visible on the barrel of the 7.5cm gun reads 'Braun-Ark' and refers to the fluid used in the recoil mechanism.

From August 1944 a number of independent Panzer brigades were raised and for the most part these formations were short-lived. As the combat histories of these brigades are covered in *TankCraft 24, Panther Tanks: German Army Panzer Brigades, Western and Eastern Fronts, 1944-1945* I have only mentioned the elements which operated the Sdkfz 251/9 briefly here. Note that although each of the brigades were authorised a number of 7.5cm-armed halftracks, many were never delivered and where exact numbers are known I have tried to make this clear.

Panzer-Brigade 101. The first ten brigades, numbered 101 to 110, contained a Panzergrenadier battalion made up of five companies. The two rifle companies each had two Sdkfz 251/9 halftracks while the fifth heavy company had six vehicles. Panzer-Brigade 101 fought in Estonia in September and October 1944 with Panzerverband Strachwitz and Panzer-Brigade Gross where the Sdkfz 251/9 halftracks are mentioned.

Panzer-Brigade 102. The brigade's Panzergrenadier-Bataillon 2102 was allocated ten Sdkfz 251/9 halftracks and it appears that these had all been received by the time the brigade left for the front in August 1944. Unusually this formation had more vehicles than qualified drivers and several pieces of heavy equipment had to be left in Germany. Several photographs of the brigade's Sdkfz 251 halftracks show them painted in what appears to be a factory-applied camouflage scheme.

Panzer-Brigade 103. The brigade's Panzergrenadier-Bataillon 2103 was allocated ten Sdkfz 251/9 halftracks and it appears that these had all been received. From August 1944 the brigade took part in the battles in Latvia and East Prussia and on 27 October 1944 was disbanded with most of the surviving elements of the Panzergrenadier battalion going to 5.Panzer-Division.

The Stab or brigade headquarters was used to command a number of ad-hoc formations and in January 1945 was incorporated in Panzer-Regiment z.b.V. Coburg (1). The regiment's infantry and reconnaissance units were allocated a total of eight Sdkfz 251/9 halftracks. This formation is sometimes referred to as Kampfgruppe Panzer-Brigade 103 but it should not be considered a rebuilt version of the original formation.

In March 1945 the remnants were used to build Panzer-Division Müncheberg but the halftracks are not mentioned again and they may have been lost in the fighting for the Oder bridgehead or the relief of Lauban.

Panzer-Brigade 104. Formed on 15 August 1944, the brigade was engaged in the battles along the Lithuanian border and in northern Poland. The brigade's Panzergrenadier-Bataillon 2104 was allocated ten Sdkfz 251/9 halftracks but I have been unable to confirm if these were delivered. On 2 November 1944, after suffering heavy losses, the brigade was disbanded.

Notes

1. This unit eventually controlled the remnants of three Panzer battalions together with Panzergrenadier-Bataillon Neuhammer, Panzergrenadier-Bataillon Kamenz and Panzer-Aufklärungs-Abteilung Hirschberg.

Panzer-Brigade 105. The brigade's Panzer Grenadier-Bataillon 2105 was allocated a total of ten Sdkfz 251/9 halftracks but it appears none were delivered. On 23 September 1944, after losing half the Panzergrenadier battalion, the brigade was disbanded and most of the survivors were integrated into the units of 9.Panzer-Division.

Panzer-Brigade 106. Raised in late August 1944, this formation's Panzergrenadier-Bataillon 2106 was allocated ten Sdkfz 251/9 halftracks and photographic evidence shows that at least some were delivered. They appear to have been on hand in September and October but are not mentioned in a December report. On 14 April 1945 the surviving elements were incorporated into the newly raised Panzer-Division Clausewitz.

Panzer-Brigade 107. It would appear that Panzergrenadier-Bataillon 2107 had its full complement of ten Sdkfz 251/9 halftracks during the fighting in Holland in September 1944. By the end of October 1944 the brigade had been reduced to a weak battle group and on 9 November was officially disbanded.

Panzer-Brigade 108. Formed on 15 September 1944, the brigade's Panzergrenadier-Bataillon 2108 was authorised ten Sdkfz 251/9 halftracks. A report written by the brigade commander and dated 9 September 1944 specifically mentions these vehicles. The brigade was formally disbanded on 30 October 1944 after the fighting for Aachen.

Panzer-Brigade 109. Formed in late September 1944, the brigade's Panzergrenadier-Bataillon 2109 was allocated ten Sdkfz 251/9 halftracks but I have not been able to confirm that these were delivered. On 5 November 1944 the remnants of Panzer-Brigade 109 were formally incorporated into Panzer-Division Feldherrnhalle

Panzer-Brigade 110. On 28 September 1944, just three days after the brigade's organisation was completed, the brigade commander, Oberstleutnant Wilhelm Schöning, was advised that Panzer-Brigade 110 was to be disbanded.

Panzer-Brigade 111. The last three brigades raised were built with full Panzer and Panzergrenadier regiments and were in fact weak divisions without an artillery element. Panzer-Brigade 111 was formed in early September and took part in the battles in Alsace-Lorraine. Sources differ on the allocation of Sdkfz 251/9 halftracks with one stating that two were on hand with the first company of Panzergrenadier-Regiment 2111 while others record just one vehicle with Panzer-Aufklärungs-Kompanie 2111. On 1 October 1944 the brigade was officially dissolved.

Panzer-Brigade 112. The brigade's Panzergrenadier-Regiment 2112 had very few, if any, armoured halftracks other than a number of Sdkfz 251/21 Drilling which equipped the regiment's Flak-Kompanie. At least one sources states that the regiment's antitank company contained four Sdkfz 251/22 halftracks but this cannot be correct as the first of these vehicles had not yet left the assembly lines when the brigade was disbanded in late September 1944.

Panzer-Brigade 113. Formed in early September 1944, the brigade's Panzergrenadier-Regiment 2113 had no halftracks and the regiment's support platoon was equipped with towed anti-tank guns. The brigade's Aufklärungs-Kompanie 2113 reported that eight Sdkfz 251 halftracks were on hand but none were the 7.5cm-armed version. In early October 1944 the brigade was disbanded.

Fuhrer-Begleit-Brigade. The brigade served on the Eastern Front and later took part in the Ardennes Offensive, reporting on 30 December that ten Sdkfz 251/9 halftracks were on hand with I. Abteilung, Panzergrenadier-Regiment FBB. On the same day four Sdkfz 251/22 vehicles were allocated and these arrived at some time in the following month. On 26 January 1945 the brigade was reformed as the Führer-Begleit-Division and fought in the battles along the Vistula and was effectively destroyed in the fighting around Spremberg in April although small Kamfgruppen continued to operate in northern Czechoslovakia. An Sdkfz 251/9 halftrack of the Führer-Begleit-Division is shown on page 28 of the Camouflage & Markings section.

Fuhrer-Grenadier-Brigade. Initially formed on 10 July 1944 as Kampfgruppe Führer-Grenadier-Brigade (FGB), this formation was made up from elements of Ersatz-Brigade Grossdeutschland and men of the Führer-Grenadier-Battalion, a unit of Hitler's bodyguard. In October 1944 the brigade took part in the recapture of Gummbinen in East Prussia and was transferred to the West to participate in the Ardennes Offensive. On 30 December the brigade reported that eleven Sdkfz 251/9 halftracks were on hand with I.Bataillon, Panzergrenadier-Regiment FBB (1).

The brigade was raised to division status on 26 January 1945 and took part in the fighting in Pomerania in February 1945 and in the attempts to retake Lauban in the following month.

In April 1945 the division was engaged in the defensive battles around Vienna and eventually surrendered to units of the US Army at Linz who promptly handed their prisoners over to the Soviets.

Notes

1. In some sources this battalion is referred to as I.(Füsilier) Bataillon but the published list of Feldpostnummern for this formation suggests that this is not correct.

PANZER-BRIGADE 107

The creation of the independent Panzer brigades, which began in July 1944, was so rushed that our knowledge of the actual establishments is quite limited. Shown here is Panzergrenadier-Bataillon 2107, the infantry component of Panzer-Brigade 107, one of the few formations for which we have detailed information. Each brigade also contained Panzer, engineer, signals and reconnaissance elements and these are examined in TankCraft 24, Panther Tanks: German Army Panzer Brigades, Western and Eastern Fronts, 1944-1945. *It is not certain that Panzergrenadier-Bataillon 2107 received its allocation of Sdkfz 2 Kettenrad halftrack-motorcycles and there is no mention of them in the lists of vehicle deliveries. The brigade headquarters certainly never received any of its authorised number and was instead equipped with civilian cars. A total of 157 Sonderpanzerwagen were delivered by the time the brigade left Truppenübungsplatz Mielau for the Western Front which would suggests that all the halftracks shown here were actually on hand when Panzergrenadier-Bataillon 2107 went into combat. The organisation shown here should be compared with the official Kriegstärkenachweisung (KstN).*

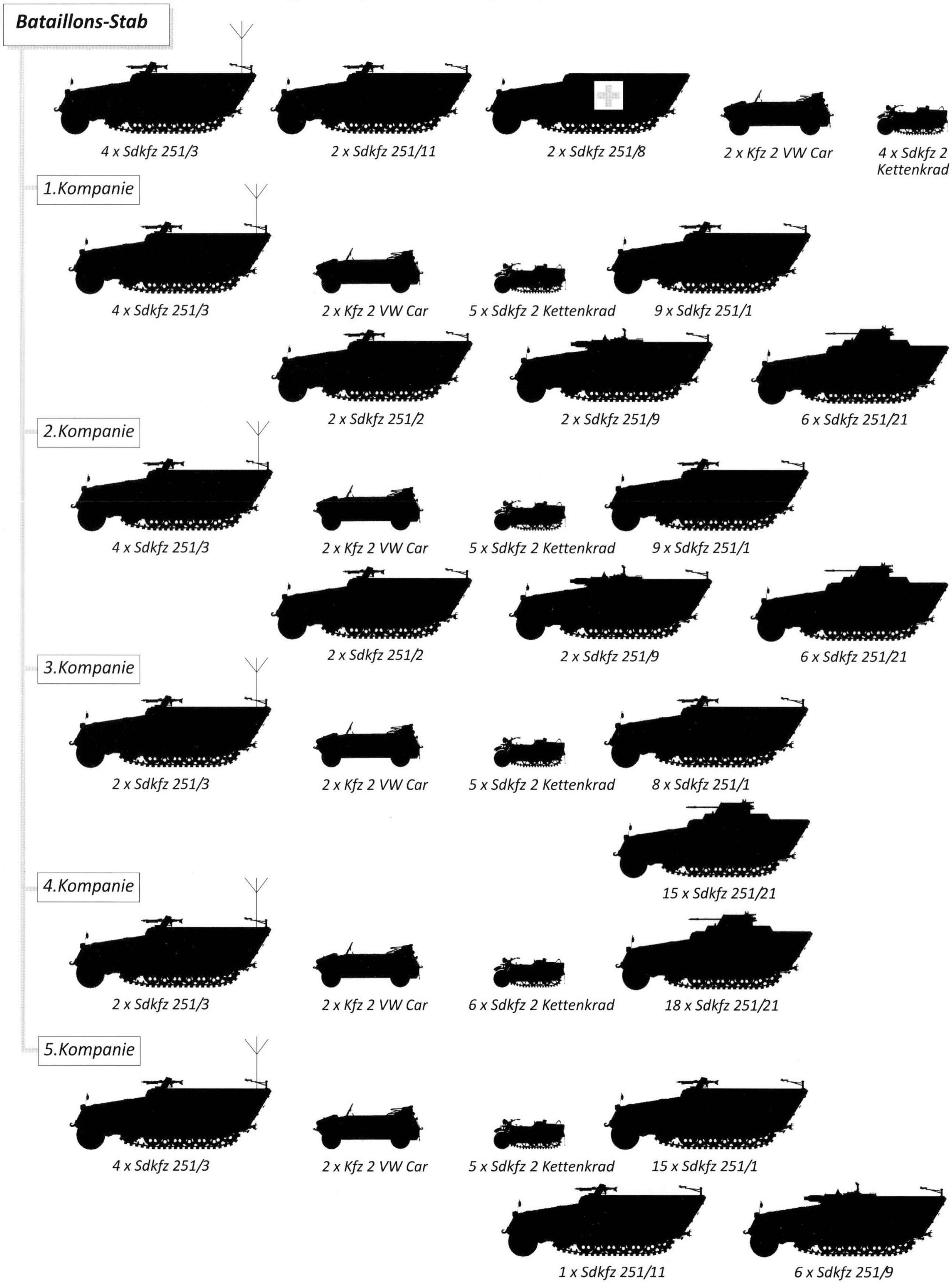

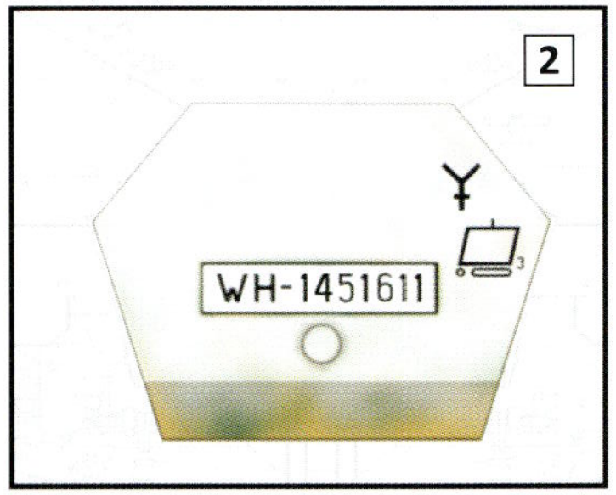

2

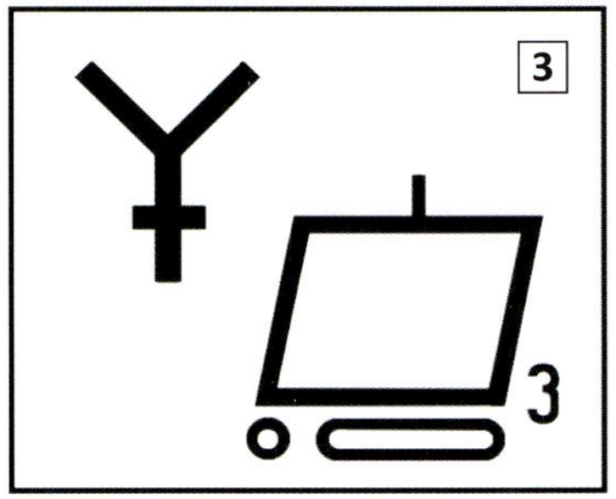

3

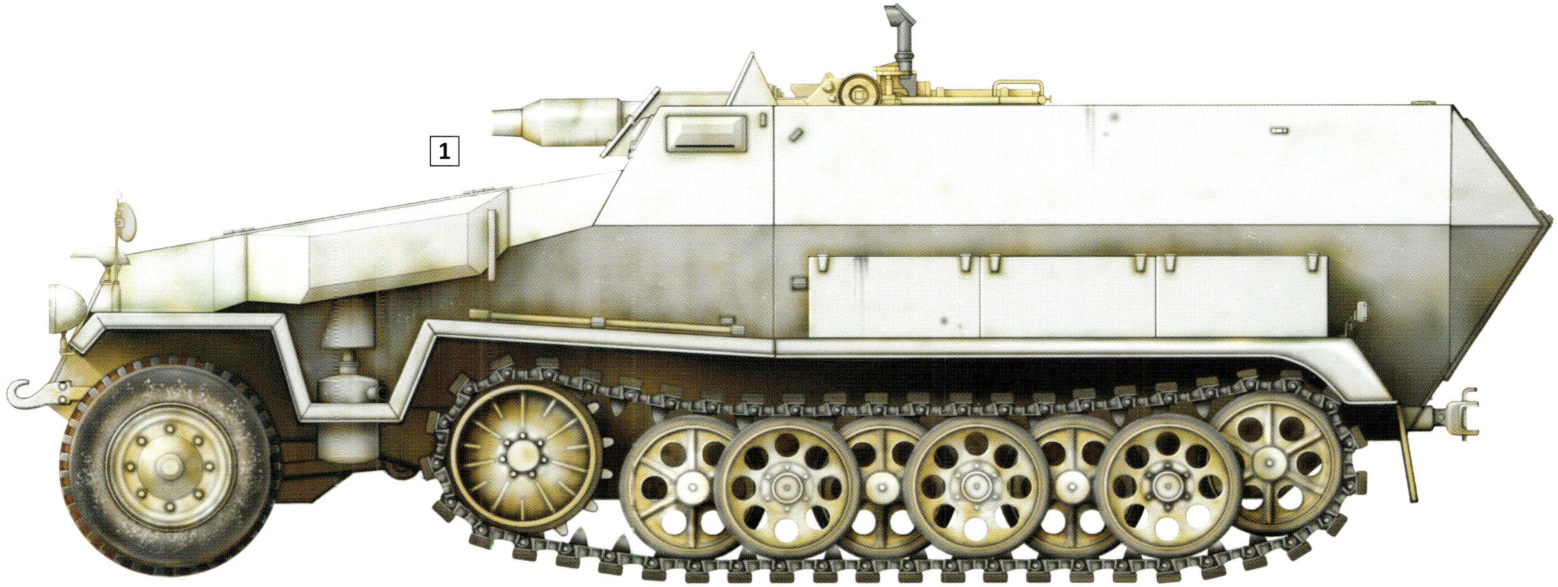

1

1. Sdkfz 251/9. 16.Panzer-Division. Russia, winter 1944-1945. Photographed with other vehicles of the division's reconnaissance battalion, this halftrack is completely coated in winter whitewash camouflage. The markings shown here are based on those carried by the other halftracks in our photograph and comprise the division's unit insignia and the tactical symbol denoting an Aufklärungs company (2 and 3).

4. Sdkfz 251/9. 5.SS-Panzer-Division. Poland, summer 1944. The III.Battalion of the division's SS-Panzergrenadier-Regiment 9 Germania was equipped with a number of these 7.5cm-armed halftracks and a detailed order of battle is shown on page 53. The vehicles were numbered as depicted here with the numbers of 9.Kompanie beginning in 1, 10.Kompanie with 2, 11.Kompanie with 3 (5) and 12.Kompanie with 4. At least one halftrack displayed kill markings (6).

6

5

4

1

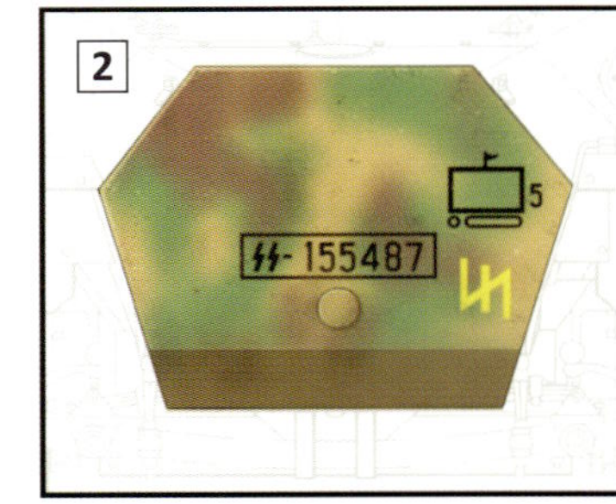

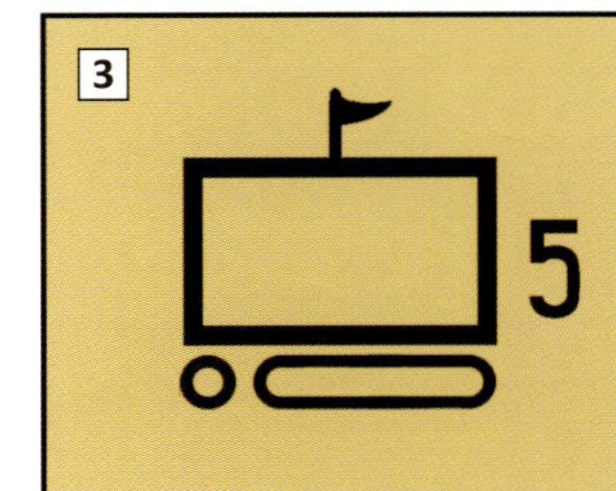

1. Sdkfz 251/9. 2.SS-Panzer-Division. France, summer 1944. Attached to 5.Kompanie of the division's SS-Auklärungs-Abteilung 5, this halftrack displays several interesting features including the metal rail welded to the hull sides above the roadwheels and, not visible in our illustration, steel-cleat tracks. Note also the Scherenfernrohr, or scissor binoculars, behind the 7.5cm gun. The markings of the division and 5.Kompanie are shown at right (2 and 3).

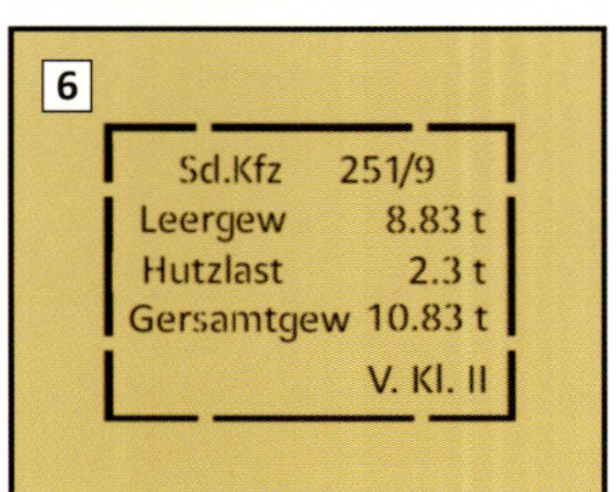

4. Sdkfz 251/9. 2.SS-Panzer-Division. France, July 1944. The III.Bataillon of the division's SS-Panzergrenadier-Regiment was fully armoured and its vehicles were numbered as shown here, with company number followed by what may be the platoon and individual vehicle or simply a consecutive number. The unusual Balkenkreuz is an identifying feature of this battalion's vehicles. All carried the division's unit insignia (5) and many had the factory-applied loading label (6).

4

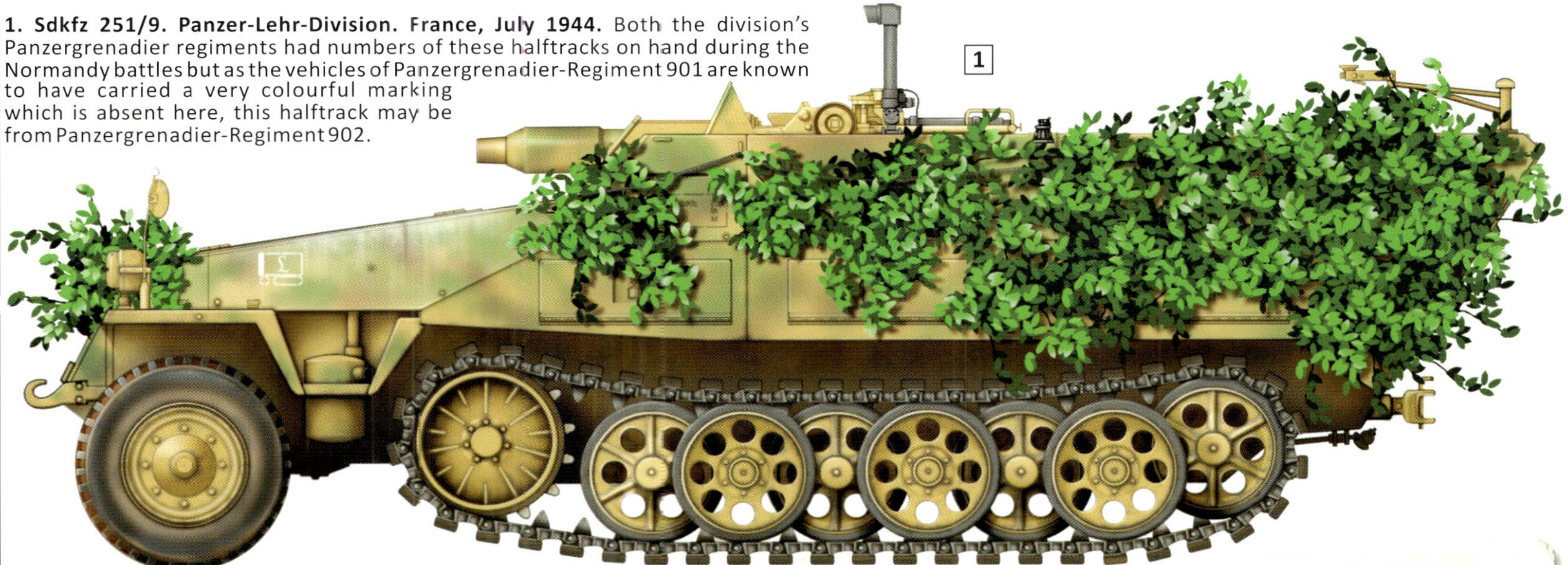

1. Sdkfz 251/9. Panzer-Lehr-Division. France, July 1944. Both the division's Panzergrenadier regiments had numbers of these halftracks on hand during the Normandy battles but as the vehicles of Panzergrenadier-Regiment 901 are known to have carried a very colourful marking which is absent here, this halftrack may be from Panzergrenadier-Regiment 902.

Above: Note the tactical marking, which also contains the division's unit insignia, on the hull side. This is just visible in our photograph (2) and is also shown in an enlarged view (3).

4. Sdkfz 251/9. 12.SS-Panzer-Division. France, June 1944. The III.Bataillon of the division's SS-Panzergrenadier-Regiment 26 was organised as a Gepanzert infantry battalion and the vehicles were numbered as shown here. The system was identical to that described for SS-Panzergrenadier-Regiment 9 on page 17, identifying this halftrack as a 10.Kompanie Sdkfz 251/9. The style of numbering can be seen in our photograph (5) where the subject of this profile is in fact the third vehicle from the right. Most carried the division's unit insignia (6) on the front and rear hull.

2

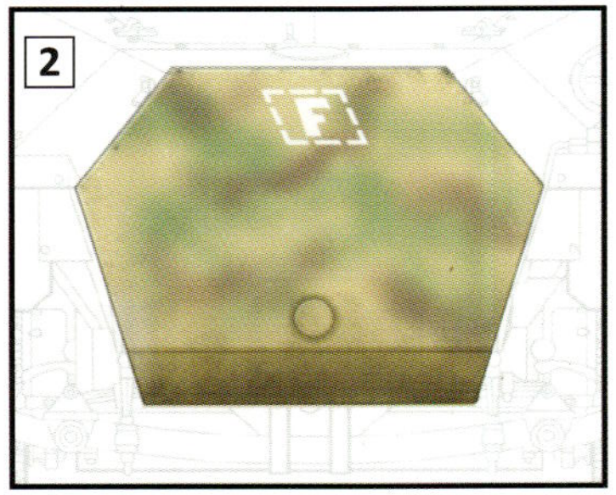

3

1. Sdkfz 251/9. 10. SS-Panzer-Division. France, September 1944. The first battalion of the division's SS-Panzergrenadier-Regiment 21 was organised as a Gepanzert formation and contained companies 1 to 4. The halftracks were marked as shown here, with the company number and a version of the division's unit insignia, sometimes also carried on the hull front (2).

1

Left: Variations of the division's unit insignia existed but most featured a stylised letter F enclosed in a Panzer rhomboid (3). A photograph of this halftrack is reproduced on page 54.

4. Sdkfz 251/9. Unidentified unit. France, August 1944. This vehicle is of interest in that it is based on the earlier Sdkfz 251 ausf C model and has been modified with additional armour to better protect the crew on the hull sides around the top. Although the company number 523 is rendered in the same style and position as that seen on halftracks of Panzer-Aufklärungs-Abteilung 2 any positive identification is not possible. Shown at right are examples of the ammunition used with the 7.5cm L/24 gun including armour-piercing (5), high explosive (6) and smoke (7).

4

5 6 7

1

2

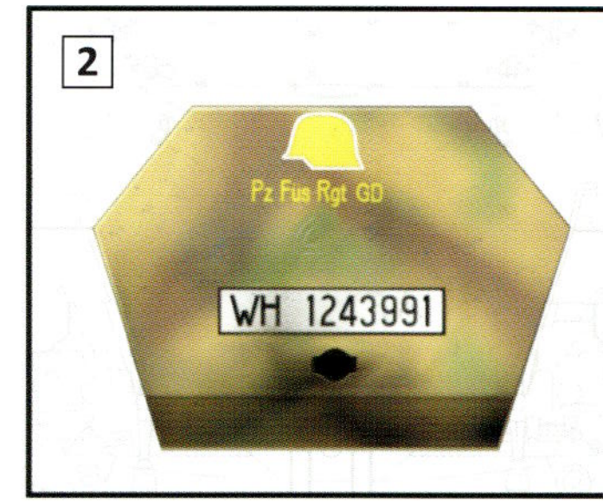

3

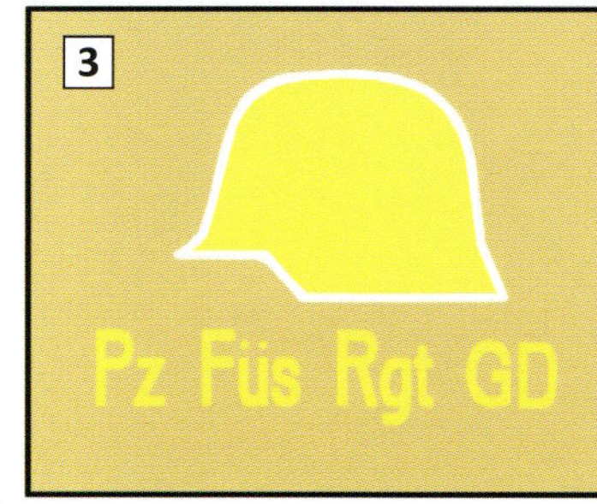

1. Sdkfz 251/9. Panzergrenadier-Division Grossdeutschland. East Prussia, spring 1945. This halftrack is fitted with the later mount for the 7.5cm gun which went into production in late December 1943. The division's famous unit insignia is marked prominently on the hull front (2) and seems to have been painted over a number 2. The notation (3) identifies the division's Panzerfusiler regiment and although it's rendered here in yellow it could have been another similar shade. A photograph of this vehicle is shown on page 11.

5

4

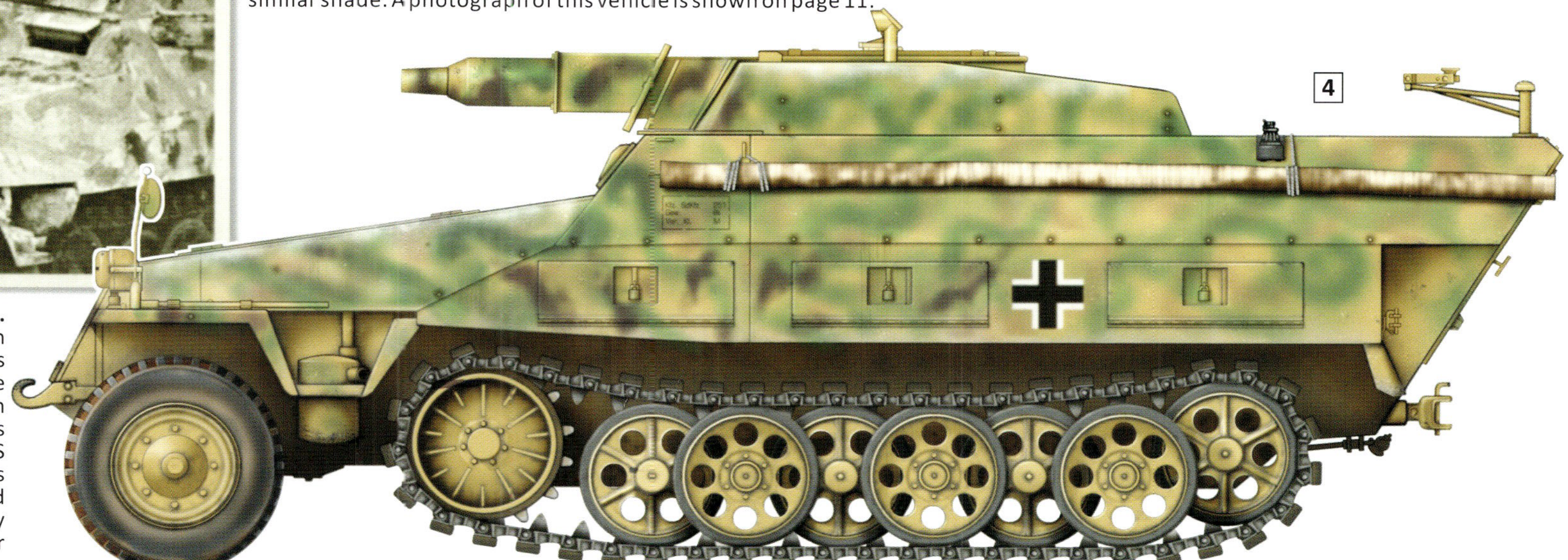

4. Sdkfz 251/9. Unidentified unit. Poland, spring 1944. Said to have been photographed during the battles around Tarnopol in March 1944, the uniform details of the crew, visible in other images, confirms that this halftrack was allocated to a Waffen-SS unit. The log carried on the hull side was used to extricate vehicles that had become bogged down. The roughly applied camouflage is evident in our photograph (5).

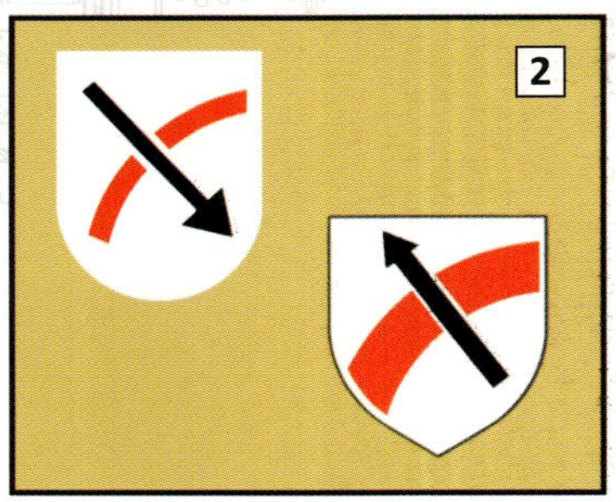

1. Sdkfz 251/9. 6.Panzer-Division. Poland, autumn 1944. In August 1944 armoured vehicle manufacturers were ordered to paint vehicles in camouflage colours before they left the factories. Although the exact date when this order was first implemented is not known, as many as 330 halftracks of this type may have been painted in factory-applied schemes before production ended. This vehicle was allocated to 8.Kompanie of the division's Panzergrenadier-Regiment 114 and its association with 20.Panzer-Division by some sources is incorrect. Variations of the unit insignia existed (2) and the examples shown here were seen on the PzkpfwV Panther tanks of the division's Panzer-Regiment 11 in late 1944.

3. Sdkfz 251/9. Panzergrenadier-Division Brandenburg. Eastern Front, spring 1945. The side view in our illustration is completely conjectural as only the hull front is visible in the photograph on which the profile is based. The division's colourful unit insignia was carried on the hull front plate (4) and was based on the Grossdeutschland division's Stalhelm design with the Brandenburg eagle, almost certainly in red, superimposed (5).

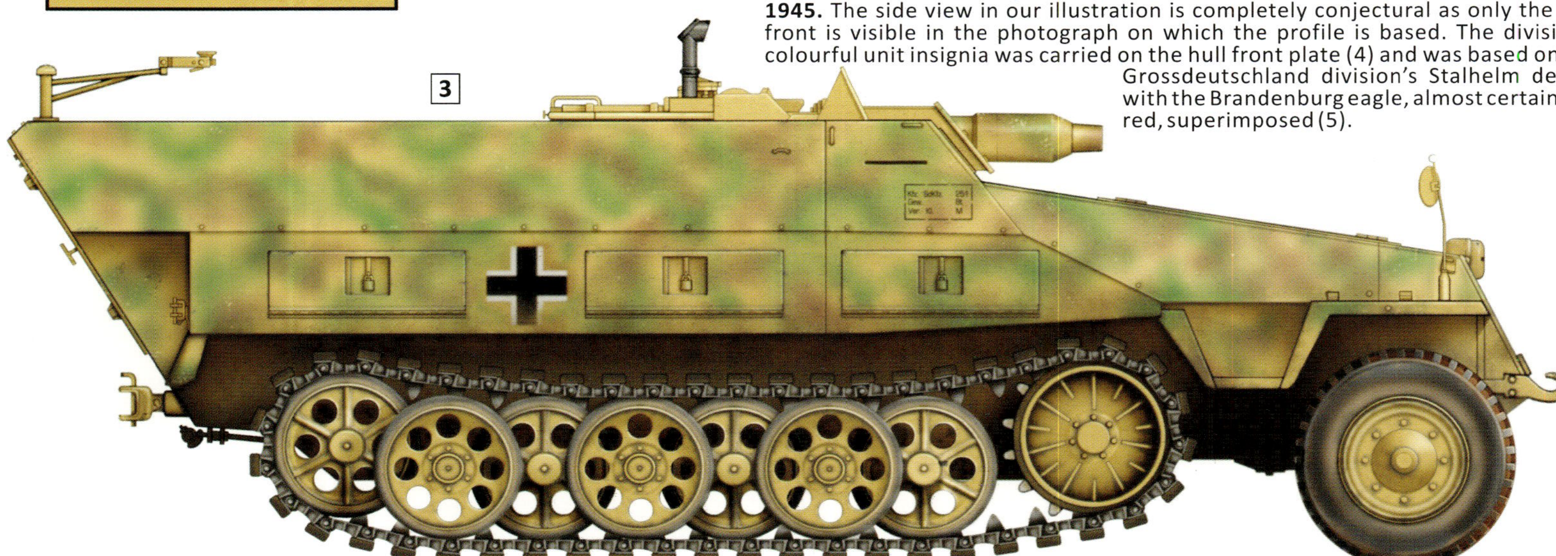

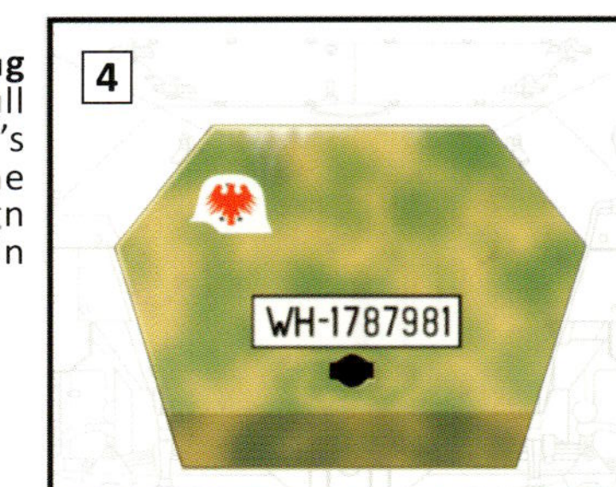

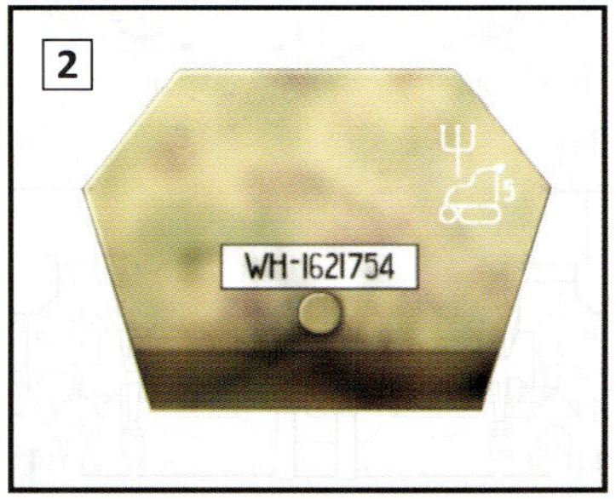

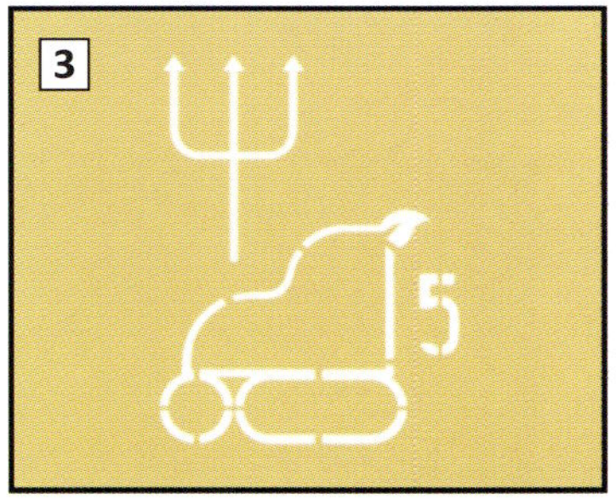

1. Sdkfz 251/9. 2.Panzer-Division. Belgium, December 1944. Captured intact after the Ardennes Offensive, this halftrack from the division's Panzer-Aufklärungs-Abteilung 2 may have been one of the vehicles of Kampfgruppe von Böhm which managed to reach the Meuse River crossings near Dinant before being effectively destroyed on Christmas Day 1944. The unit markings are not discernable in the photograph on which this illustration is based and those used in Normandy are shown here (2) and identify the fifth, armoured car, company (3).

4. Sdkfz 251/9. 1.SS-Panzer-Division. Belgium, December 1944. Although it is difficult to be certain, this halftrack was probably one of the vehicles of III.Bataillon, SS-Panzergrenadier-Regiment 2 that was attached to Kampfgruppe Peiper during the Ardennes Offensive. Photographs of the battle group's other halftracks, including Obersturmbannführer Peiper's command vehicle, show that all seem to be painted in a factory-applied camouflage scheme. The division's unit insignia was carried on the hull front (5) and on the rear, usually on the right-hand side.

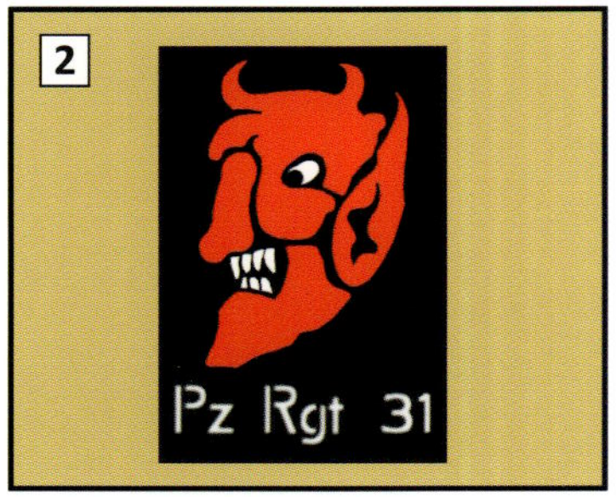

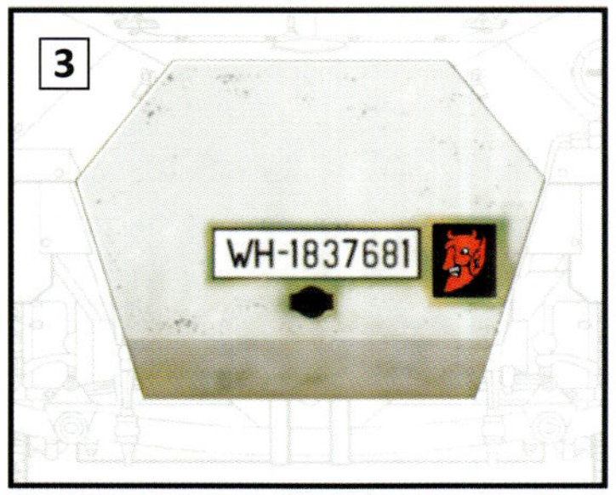

1. Sdkfz 251/9. Panzer-Regiment 31. East Prussia, winter 1944-1945. Although I cannot find any record of these vehicles being operated by the regiment, the unit marking is quite clear in the photograph on which this profile is based. The practice of clearly identifying a unit by a combination of the formation insignia and an abbreviated version of its title (2) was common by the end of 1944, particularly in the East. Other halftracks of the division are known to have displayed the formation sign on the hull front (3) and at the rear.

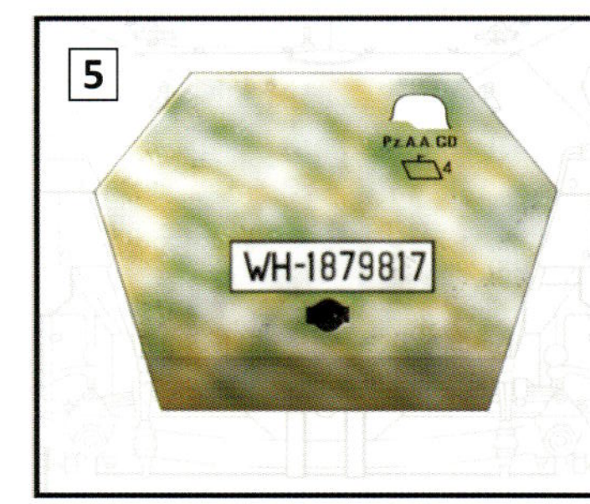

4. Sdkfz 251/9. Panzergrenadier-Division Grossdeutschland (GD). East Prussia, early 1945. Both the division's Panzergrenadier and Panzefüsilier regiments and also the Panzer-Aufklärungs battalion operated a number of these halftracks and unfortunately the 4.Kompanie number does not narrow the identification. A possible rendering of the markings for Panzer-Aufklärungs-Abteilung GD are shown above (5).

1

2

3

1. Sdkfz 251/22. 11.Panzer-Division. Western Front, spring 1945. Although badly worn, the remnants of a factory-applied camouflage scheme are still visible on parts of this halftrack. Note that the division's unit insignia (2) is carried on the hull rear and also on the side above the front wheel and this is just visible in our photograph (3). This vehicle is also shown on pages 7 and 8.

4. Sdkfz 251/22. Unidentified unit. Czechoslovakia, spring 1945. Despite considerable effort I have been unable to identify the unit to which this vehicle belongs and this is explained further on page 4. The large dark patch on the hull side may cover a marking and seems to have a number 1 painted just above it.

4

1

2

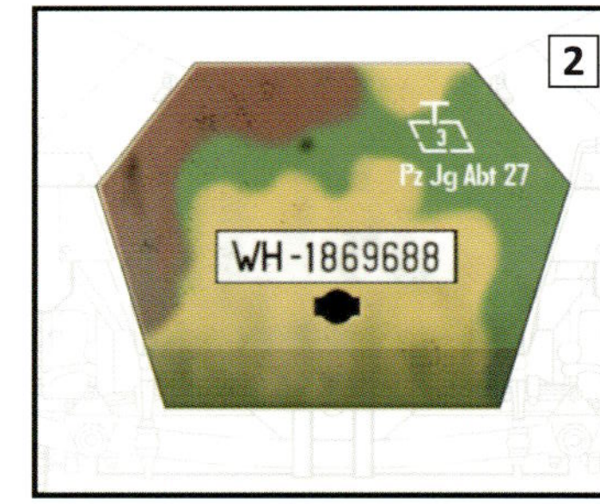

3

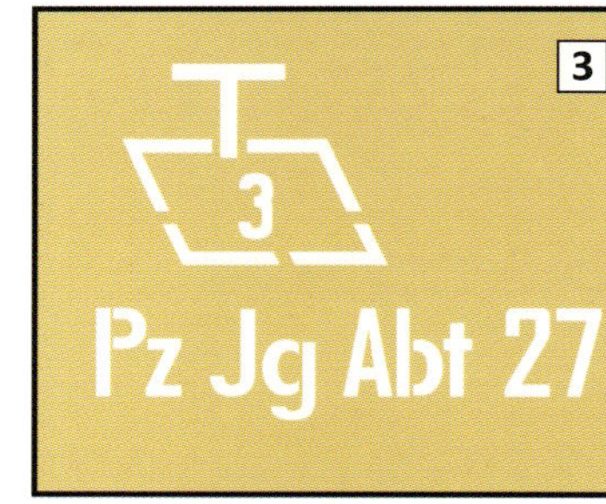

1. Sdkfz 251/22. 17.Panzer-Division. Czechoslovakia, spring 1945. The division received twelve of these 7.5cm-armed halftracks and all were allocated to Panzerjäger-Abteilung 27. The markings on the hull front (2) are discussed further on page 10 and an enlarged view is shown here (3). The company number on the hull side is rendered here in yellow but may have been a similar light shade.

4. Sdkfz 251/22. 18.Panzergrenadier-Division. Eastern Front, spring 1945. Photographed outside Berlin, this vehicle appears to be unmarked except for the registration number (5) and tactical symbol (6), which is visible but unclear in the photographed used to create this illustration. This may have been one of the halftracks that were originally issued to the short-lived Panzer-Division Schlesien.

5

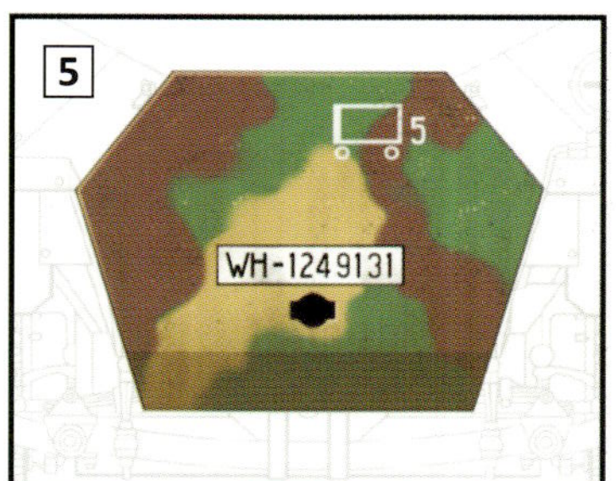

6

4

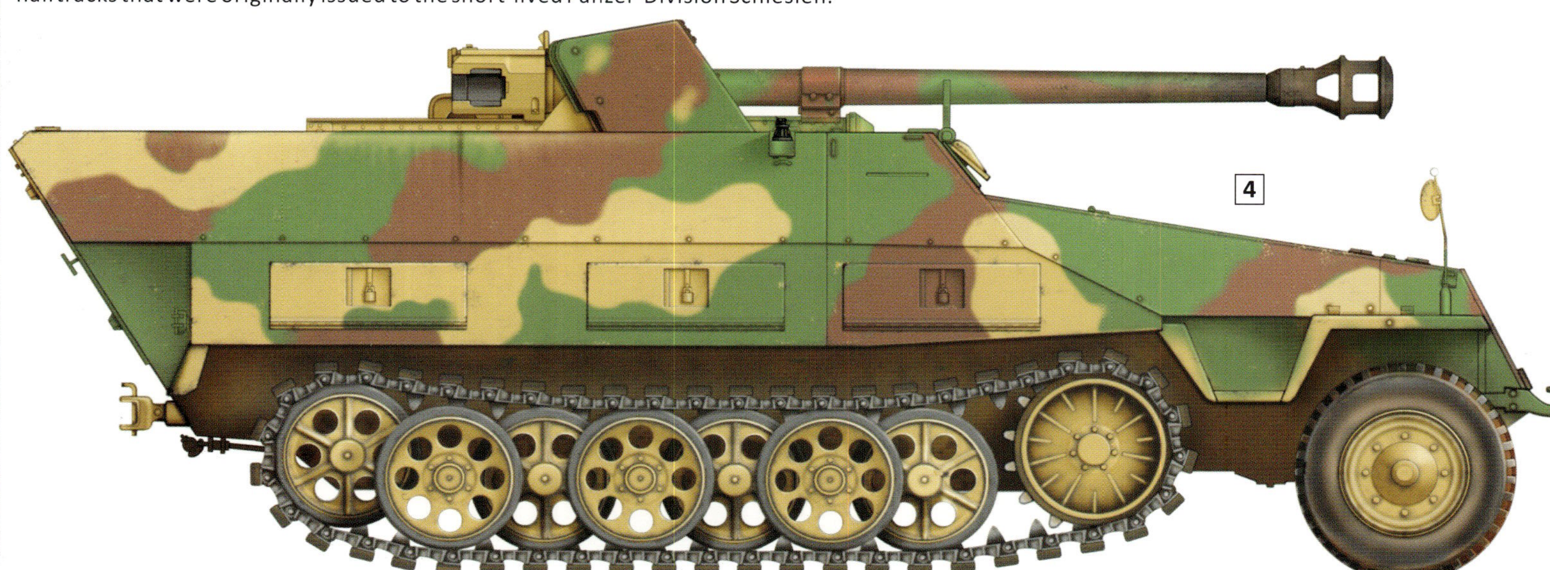

1

2
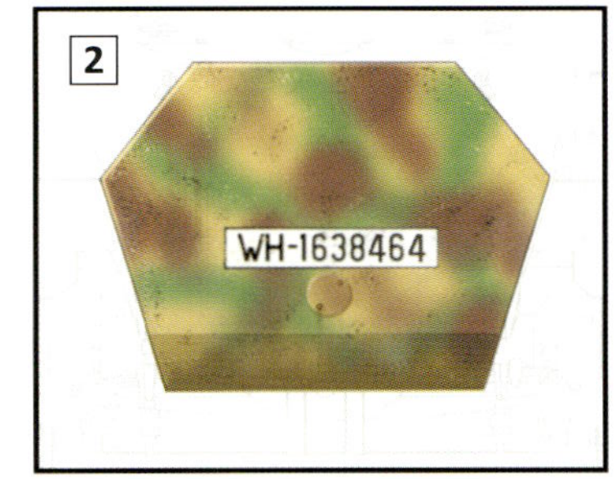

3
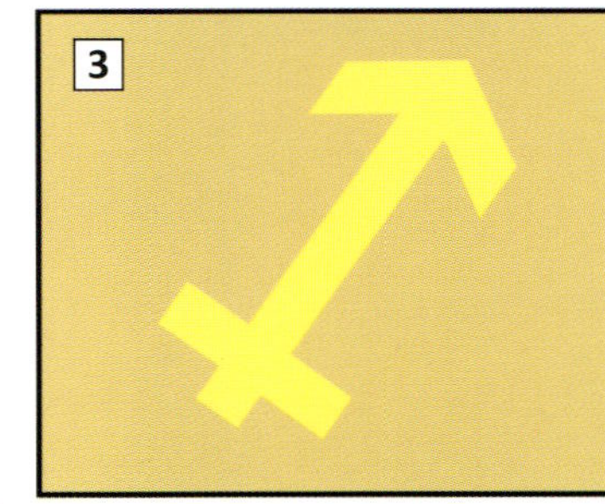

1. Sdkfz 251/9. 23.Panzer-Division. Austria, spring 1945. This vehicle is one of just two 7.5cm-armed halftracks operated by the division's Panzer-Aufklärungs-Abteilung 23. The hull front (2) carries the registration number only and the division's unit insignia (3) is shown here as a matter of interest only.

4. Sdkfz 251/22. 7.Panzer-Division. Poland, spring 1945. As most, if not all, the production models of this halftrack would have been finished in a factory-applied camouflage scheme, this vehicle may be an older version fitted with one of the conversion sets which were shipped to units in the field from December 1944. The halftrack shown here was allocated to II.Bataillon, Panzergrenadier-Regiment 6. Note that the company number is repeated on the rear door.

4
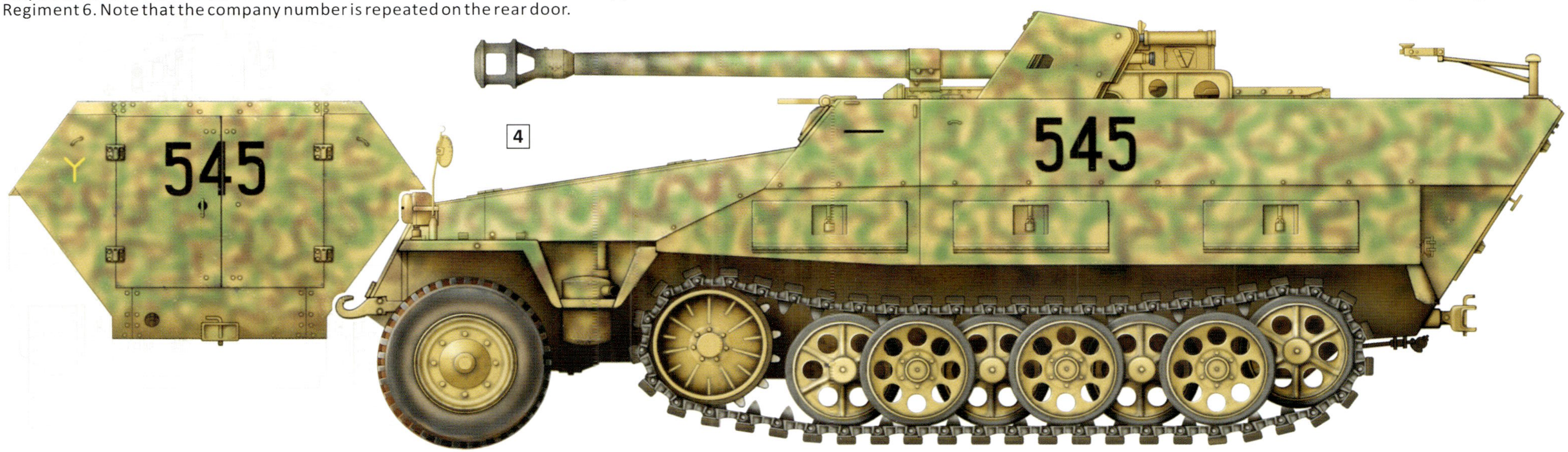

1. Sdkfz 251/9. Führer-Begleit-Division. Eastern Front, spring 1945. The unit insignia carried on the hull rear and at the front (2) may have been a variation adopted late in the war. This vehicle was probably operated by the division's Panzergrenadier-Regiment 100 which was renamed Führer-Grenadier-Regiment 1 in March 1945.

3. Sdkfz 251/22. 8.Panzer-Division. Czechoslovakia, spring 1945. This halftrack may have been allocated to the division's Panzer-Aufklärungs-Abteilung 8 or a Kampfgruppe made up from elements of the division. Note the non-standard bracket for a shovel on the side of the engine compartment. The markings on the hull front (4) are based on another vehicle photographed in the same location and include the division's unit insignia and a tactical sign denoting a Panzergrenadier company (5).

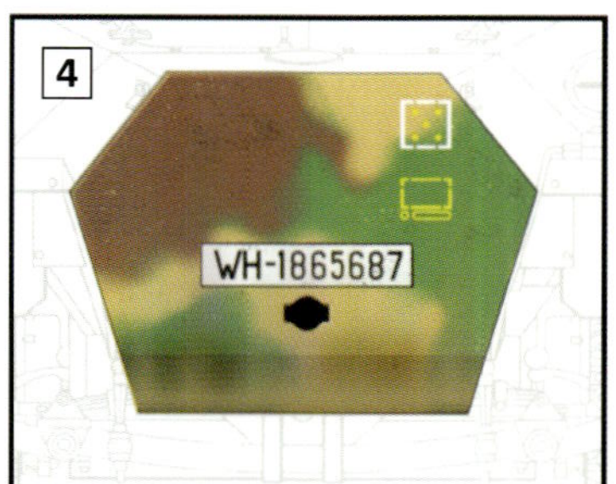

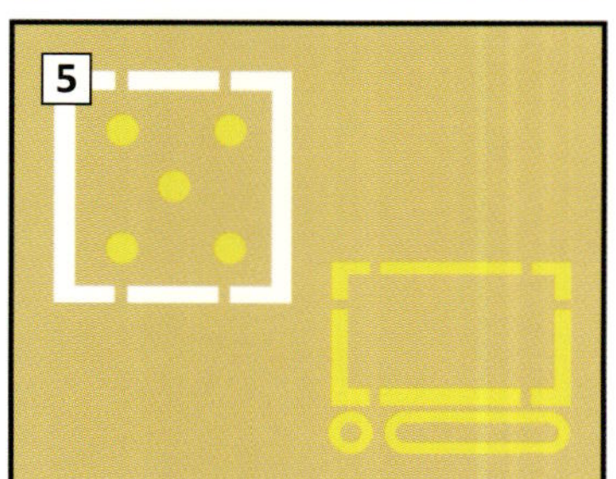

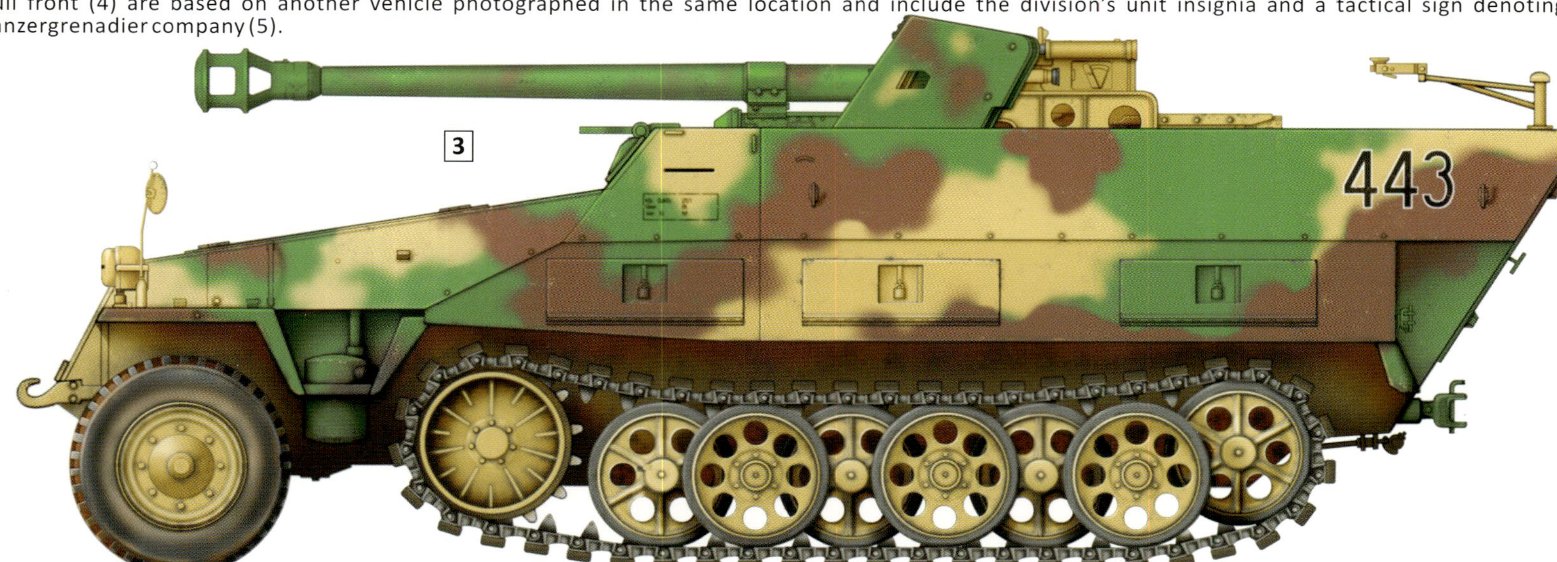

SDKFZ 251/22 PAKWAGEN

1/35 SCALE

LEO HAN

Taiwanese modeller Leo Han based his replica on the Dragon Models' 1/35 scale kit and built it without any added extras. Leo painted the completed build with Gunze Sangyo and Tamiya paints and used artist's oils for weathering.

The markings consist of the Balkenkreuz national insignia and company number and they are included in the kit. The extensive paint chipping was achieved with a fine brush. Note the missing locker door.

Another view of Leo's finished model showing the two-piece engine access doors and tool stowage on the front fenders. The field-applied camouflage, worn appearance and standard co-driver's visor are all indicative of conversions built on standard Sdkfz 251/1 halftracks.

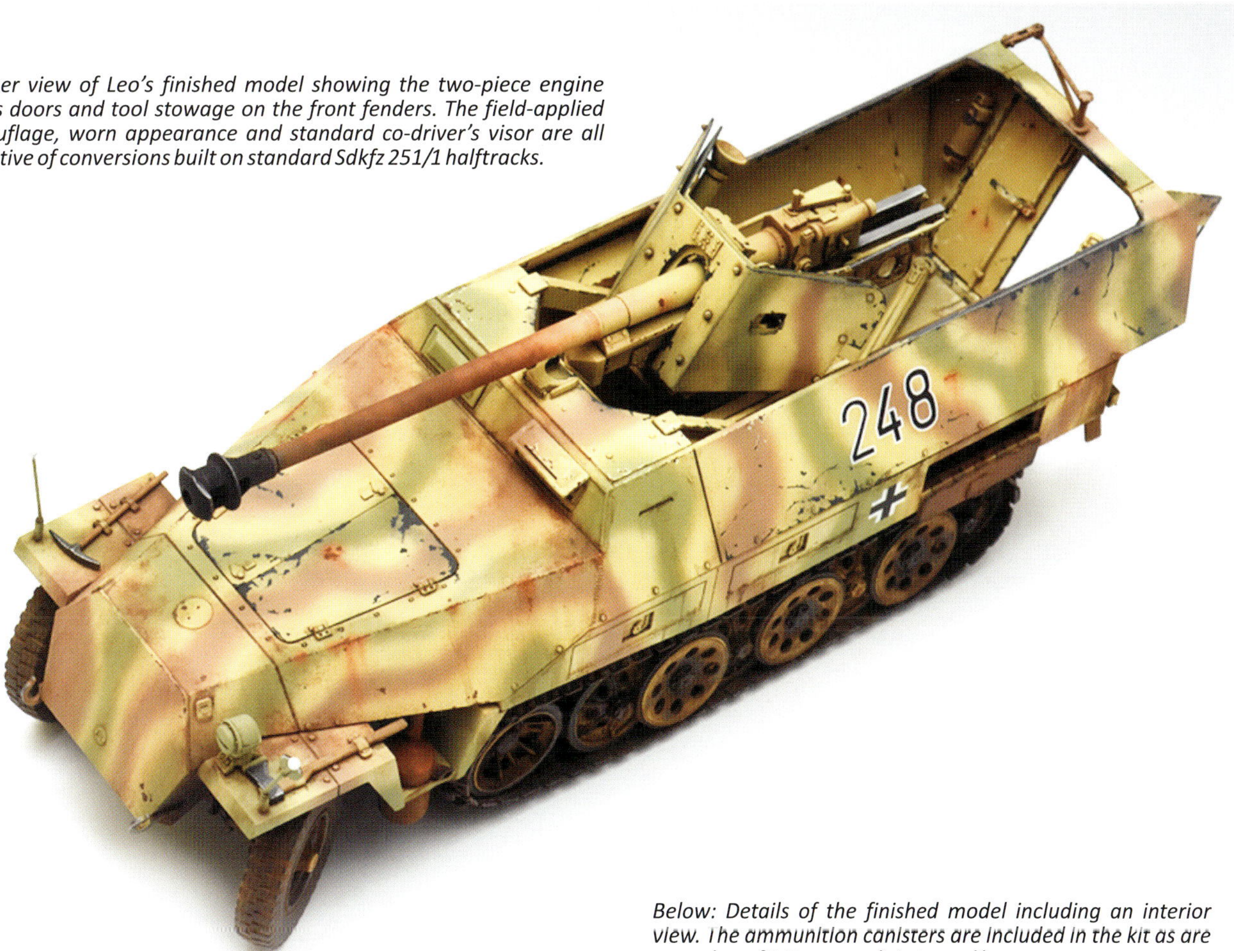

Below: Details of the finished model including an interior view. The ammunition canisters are included in the kit as are a number of 7.5cm rounds in turned brass.

SDKFZ 251/22 PAKWAGEN

1/35 SCALE

FABIO MARINI

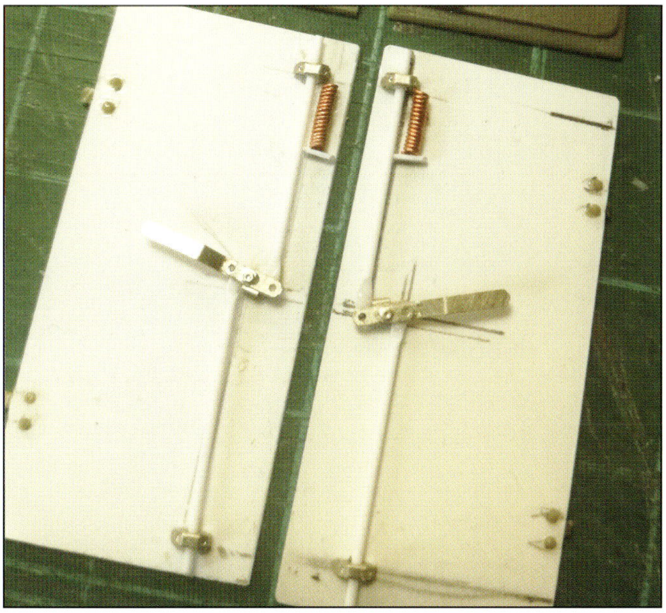

Fabio's model is based on the AFV Club 1/35 scale Sdkfz 251/22 kit. Parts of the vehicle's interior, the front fenders, the tool hooks and clasps and the locking mechanism of the rear access doors were built with photo-etched upgrade sets from Aber, Part and Voyager Model. As can be seen from the photographs, a number of small pieces were scratch-built from plastic card and rod.

The finished model was painted in Tamiya acrylics beginning with a mixture of XF-60 Dark Yellow with XF-59 Desert Yellow, lightened with a touch of XF-2 Flat White to recreate a convincing Dunkelgelb shade. A subtle camouflage pattern was added using XF-5 Flat Green and XF-64 Red Brown.

Below: 1. The front fenders built using the Aber aluminium and photo-etched brass parts. 2. The rear MG mount rebuilt with plastic rod and brass. 3. The first experiment with the number stencils. 4. The rear access doors with the company number applied.

Once the painting was complete Fabio began weathering the model with shades of green and brown oil colours and then used Dark Earth enamel to create the scratches and chipping. The markings were applied using Archer dry tansfers and stencils. Lastly the complete model was dusted with a light coat of Tamiya XF-57 Buff.

SDKFZ 251/22 PAKWAGEN

1/72 SCALE

JUANJO DOMINGUEZ

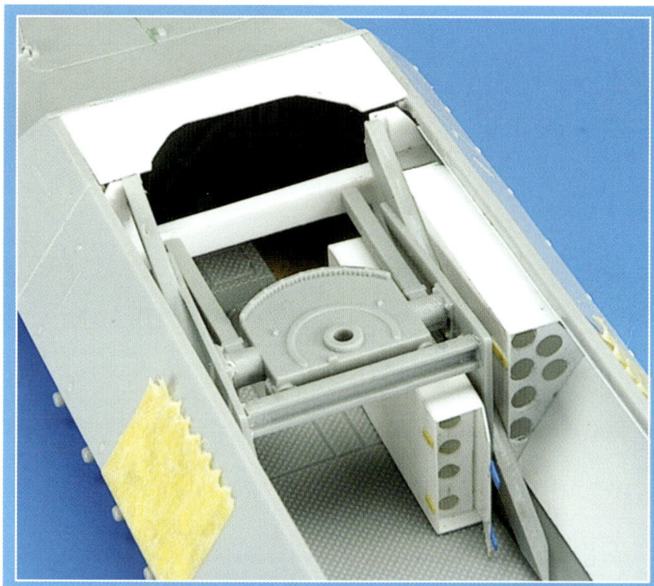

Juanjo's replica was built using the Dragon Models' 1/72 scale Sdkfz 251/22 kit with quite a bit of extra effort. He found that the photo-etched parts that accompany the kit were too thin and built replacements from plastic card. Above, at left. A view of the partially completed interior showing the steel frame that holds the 7.5cm gun and the ammunition lockers. Above, at right: Three rounds for the main gun loaded into a wooden crate. Below: the two halves of the hull ready to be painted and weathered. The photo-etched details are from Polish company Part Parts and some pieces are from other Dragon kits.

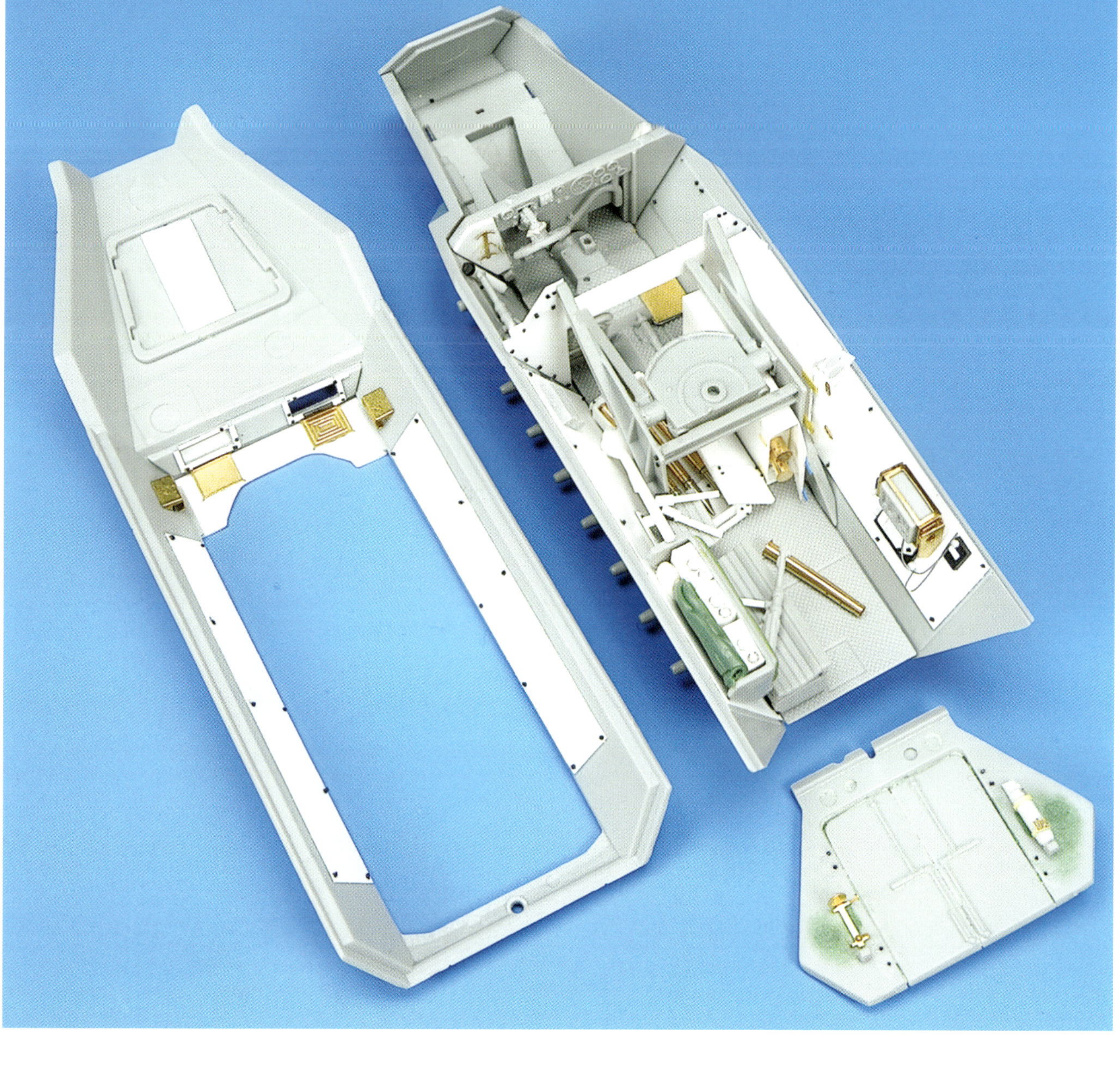

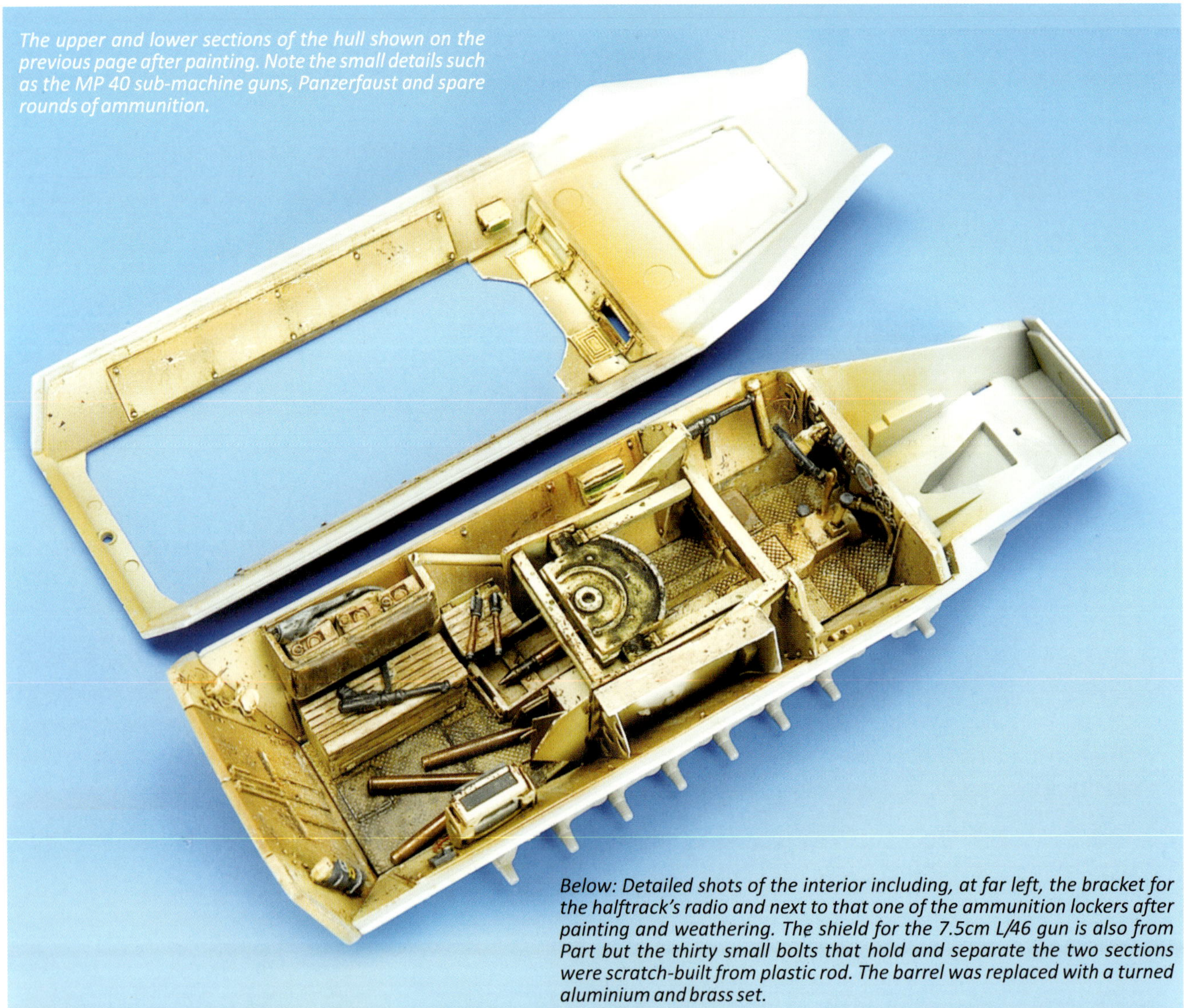

The upper and lower sections of the hull shown on the previous page after painting. Note the small details such as the MP 40 sub-machine guns, Panzerfaust and spare rounds of ammunition.

Below: Detailed shots of the interior including, at far left, the bracket for the halftrack's radio and next to that one of the ammunition lockers after painting and weathering. The shield for the 7.5cm L/46 gun is also from Part but the thirty small bolts that hold and separate the two sections were scratch-built from plastic rod. The barrel was replaced with a turned aluminium and brass set.

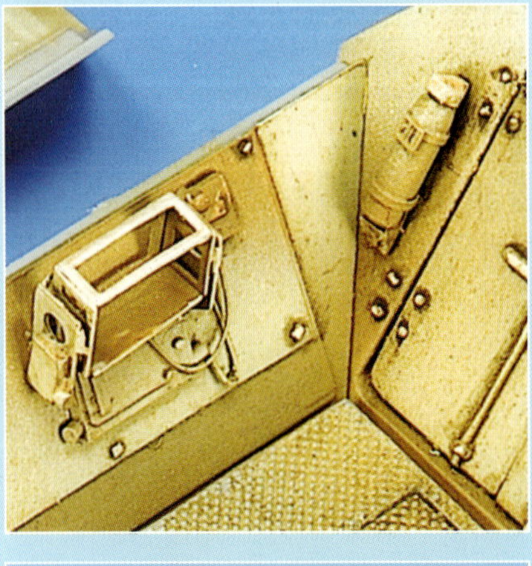

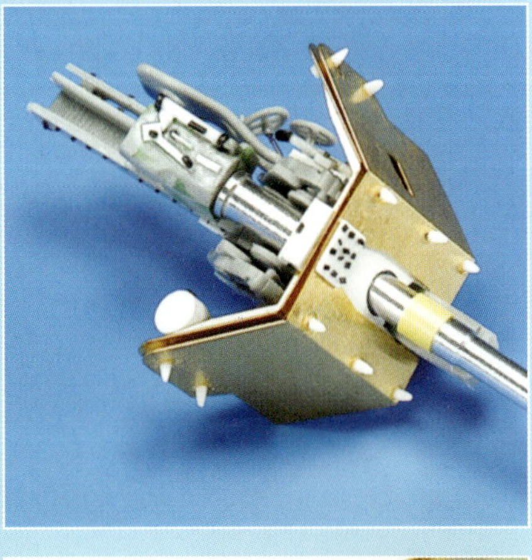

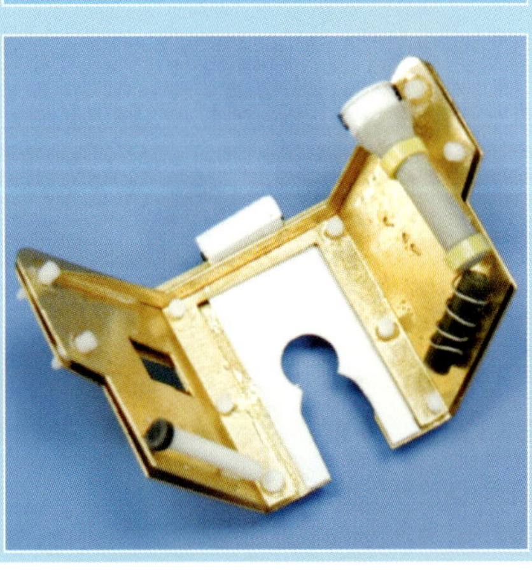

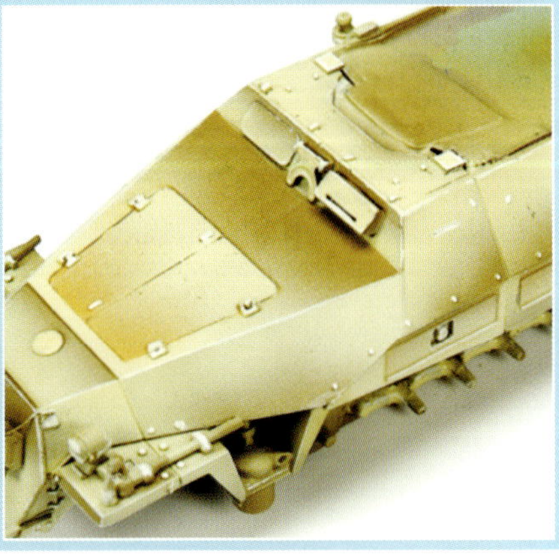

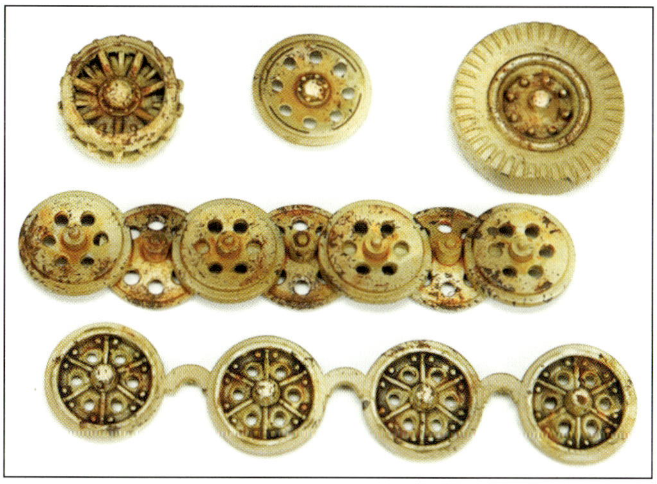

After the camouflage pattern had been applied a very delicate pin wash was added with a fine brush using a dark brown enamel. Streaks and scratches were achieved in the same way and the kit's vinyl tracks were replaced with resin pieces from OKB Grigorov. The rubber pads of the individual track links were painted one at a time.

The completed build was painted in acrylics with a coat of Dunkelgelb applied first and then Tan lightly sprayed all over. This was followed by an enamel wash in dark brown and the chipping effect was achieved with a fine brush. The photograph at top right shows just how small the model is.

SDKFZ 251/22 PAKWAGEN

1/35 SCALE

MICHELANGELO SICILIA

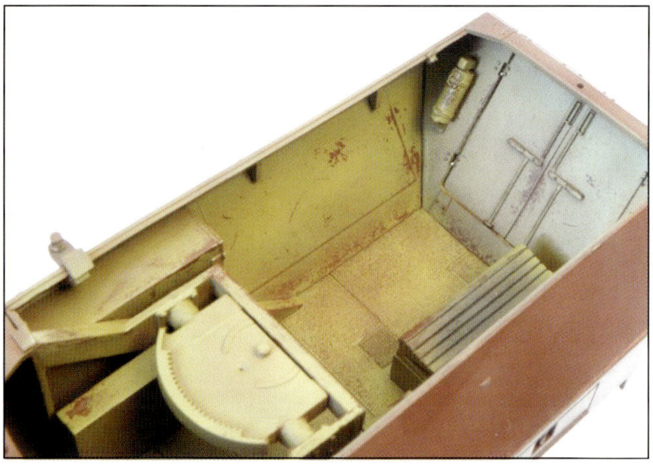

Michelangelo's model is based on the AFV Club 1/35 scale Sdkfz 251/22 Pakwagen which he was given partially built and decided to restore and complete. Once fully built the model was primed with Gunze Sangyo H3 Red Brown as a primer and then varying applications of H47 Red Brown and H69 RLM Mid Grey.

The completed model in its final camouflage scheme which was built up with layers of Gunze Sangyo H34 Cream Yellow, H47 Red Brown and H3 Red. After some experimentation with hair spray to create a chipped paint effect Michelangelo decided to repaint the model and applied the weathering at all the edges with a sponge using Gunze Sangyo H403 Dark Yellow and H3 Red.

Below: Another view of the finished model and at right, the damaged building used to create the diorama from Custom Dioramics.

As mentioned earlier, the ruined house is from Custom Dioramics. The window frame was made from scratch with Evergreen plastic pieces. The figures are from Alpine Miniatures and Platoon and were painted with Vallejo Model Colour acrylics.

Considering the important role that the Sdkfz 251 series of halftracks played in the battles of the Second World War it is surprising that a scale model of this versatile vehicle was not produced until 1972 when the Japanese toy maker Bandai released a 1/48 scale kit depicting an Sdkfz 251/1 ausf B. Within a year this was followed by Tamiya's 1/35 scale kit of an Sdkfz 251/1 ausf D and many recent Tamiya offerings share parts with this early model. Modellers who favoured the smaller scales were forced to wait until 1976 when Matchbox produced an Sdkfz 251/1 ausf B in 1/76 scale.

A number of small manufacturers produced models in resin but these were in the main quite rudimentary and often expensive. In addition, the resin was quite brittle and prone to warping during the casting process. It was not until quite recently that detailed models of the Kanonenwagen and Pakwagen became readily available and in turn this has encouraged accessory manufacturers to produce upgrade sets in photo-etched brass and other materials. Having said that, the choice of both models and accessories for the Sdkfz 251/9 and 251/22 is meagre when compared to the more glamorous types such as the Tiger and Panther, perhaps understandably so when one considers the workman-like part they played. As with the books in the TankCraft series I have concentrated here on the popular scale of 1/35 and although I had hoped to include more on the smaller scales, I decided against it when I discovered that the quality ranged from mediocre to bad. I should stress that the information presented on the following pages is far from definitive as many companies regularly withdraw products from their catalogues, often only to repackage and re-release them at some later date, and I would encourage readers to undertake their own research into the areas that interest them most. An index of manufacturers, together with suggestions for further reading, can be found on page 64 of this book.

Tamiya. The company's Sdkfz 251/9 Kanonenwagen in 1/35 scale first appeared in 1990 and at the time it was the only model of this vehicle on the market. Due primarily to its age, this kit is not as detailed as the AFV Club or Dragon Models offerings but it is a reasonably accurate rendition with an affordable price tag. One of Tamiya's most recent releases was an Sdkfz 251/1 ausf D Stuka zu Fuss which is a highly detailed and accurate replica and we may see a Pakwagen or Kanonenwagen in this scale in the near future.

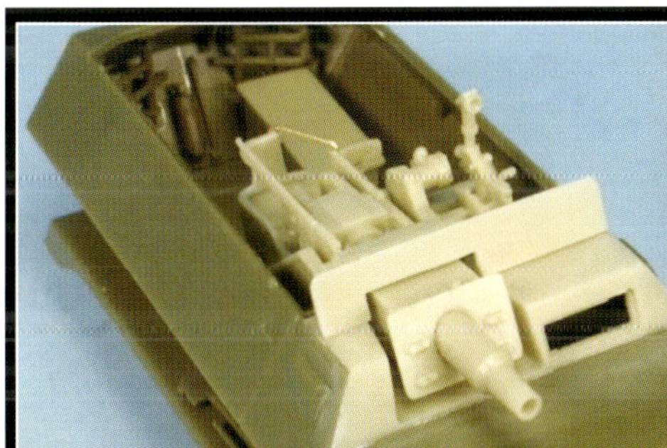

Below: Tamiya's 1/35 scale Sdkfz 251/9 Kanonenwagen built and painted according to the kit instructions. At left: The resin 7.5cm L/24 conversion set from Belgian company Gaso.Line created specifically to fit the Tamiya 1/48 scale Sdkfz 251/1 standard halftrack model.

At right: Dragon Models' 1/35 scale Sdkfz 251/22 partly built with the detail set from Eduard Model Accessories. Below: Dragon's assembled and pre-painted Sdkfz 251/22 Pakwagen kits.

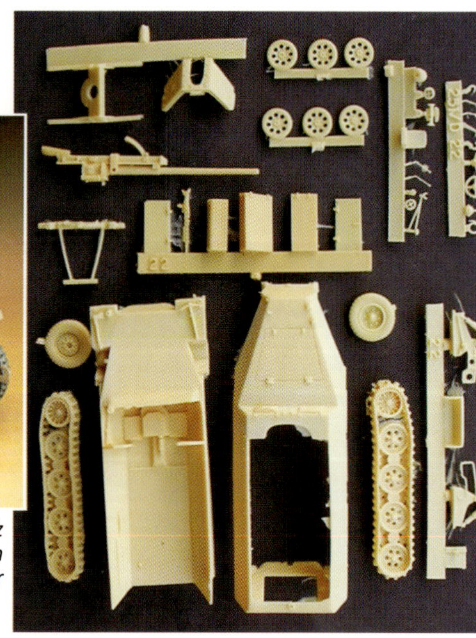

Above: The box art for Hasegawa's 1/72 scale Sdkfz 251/9 and the completed model. Left: The L/24 gun and mount for the Revell 1/72 scale kit. Right: Parts for Al-By's 1/72 scale Sdkfz 251/22 in resin.

Dragon Models. In 1/35 scale this Hong Kong-based company offers an Sdkfz 251/22 ausf D based on the company's Sdkfz 251/1 model, first released in 2004. The kit contains photo-etched brass parts and turned metal barrel and ammunition. Two sets of tracks are provided for both the early and late patterns. At the same time Dragon released a three-in-one kit which included the basic Sdkfz 251/1, the later version of the Kanonenwagen with raised armour and the Falke halftrack with night vision sights. This kit is included in the latest catalogue but the Sdkfz 251/9 is not available separately. The Sdkfz 251/22 is also available in Dragon's 1/72 scale range and is based on the company's Sdkfz 251/1 model released in 2005. The company's small scale kits are something of a benchmark and this model is no exception. Dragon Models also markets a range of assembled and pre-painted models in 1/72 scale including an Sdkfz 251/22 in two different colour schemes.

Revell. In 2010 the company released an Sdkfz 251/9 ausf C in 1/72 scale. It should be mentioned that the recent series of releases by Revell are not based on the old ESCI kits but are new models.

Al-By. This French company has been producing a range of resin models in 1/72 scale for many years, specialising in obscure vehicles. The current catalogue includes an Sdkfz 251/22, an Sdkfz 251/17 and the basic Sdkfz 251/1 halftrack.

Armourfast. Squarely aimed at the wargamer, Armourfast kits are easy and quick to build and each box contains two models. In terms of detail these models are quite basic but they are reasonably accurate and the newer offerings, produced since the company was taken over by TDL Moulding Ltd, are very attractive. The current Armourfast catalogue includes an Sdkfz 251/9 ausf C in 1/72 scale. The company also markets a range of waterslide transfers.

AFV Club. This Taiwan-based company's Sdkfz 251/9 ausf D Kanonenwagen in 1/35 scale, first released in 2004, depicts the Späte or late version of the 7.5cm armed halftrack and is based on the company's Sdkfz 251/1 kit. The barrel for the L/24 gun is rendered in turned aluminium and the mounting is very highly detailed. The single marking option depicts the vehicle of 6.Panzer-Division shown on page 22 of this book, although they are incorrectly identified. In the same year the company also released several versions of the basic halftrack including an Sdkfz 251/22 ausf D Pakwagen which utilised AFV Club's 7.5cm Pak 40 kit. Again, the main gun is in turned aluminium but the muzzle brake is injection-moulded plastic. The first edition of both models included a bonus crew figure in resin made by Hobby Fan but I have not been able to determine exactly how many kits this involved.

In 2015 AFV Club released a 1/35 scale Sdkfz 251/9 ausf C kit and in 2019 a model of the early version Sdkfz 251/9 ausf D. Interestingly, this kit was sold in France under the Best Value Models label which may have some connection to Italeri.

Hasegawa. One of the oldest plastic model companies still operating, Hasegawa produces 1/72 scale models of the later Sdkfz 251/9 and also the Sdkfz 251/22. Both are quite pleasing, easy to build and inexpensive but they are let down somewhat by poor quality of the tracks, as a number of Hasegawa kits are.

Above: The box art for AFV Club's 1/35 scale Sdkfz 251/9 ausf C and the Sdkfz 251/9 ausf D built following the kit instructions. At right: The detailed 7.5cm L/24 gun. Note the gunner's seat, elevating gears and sight.

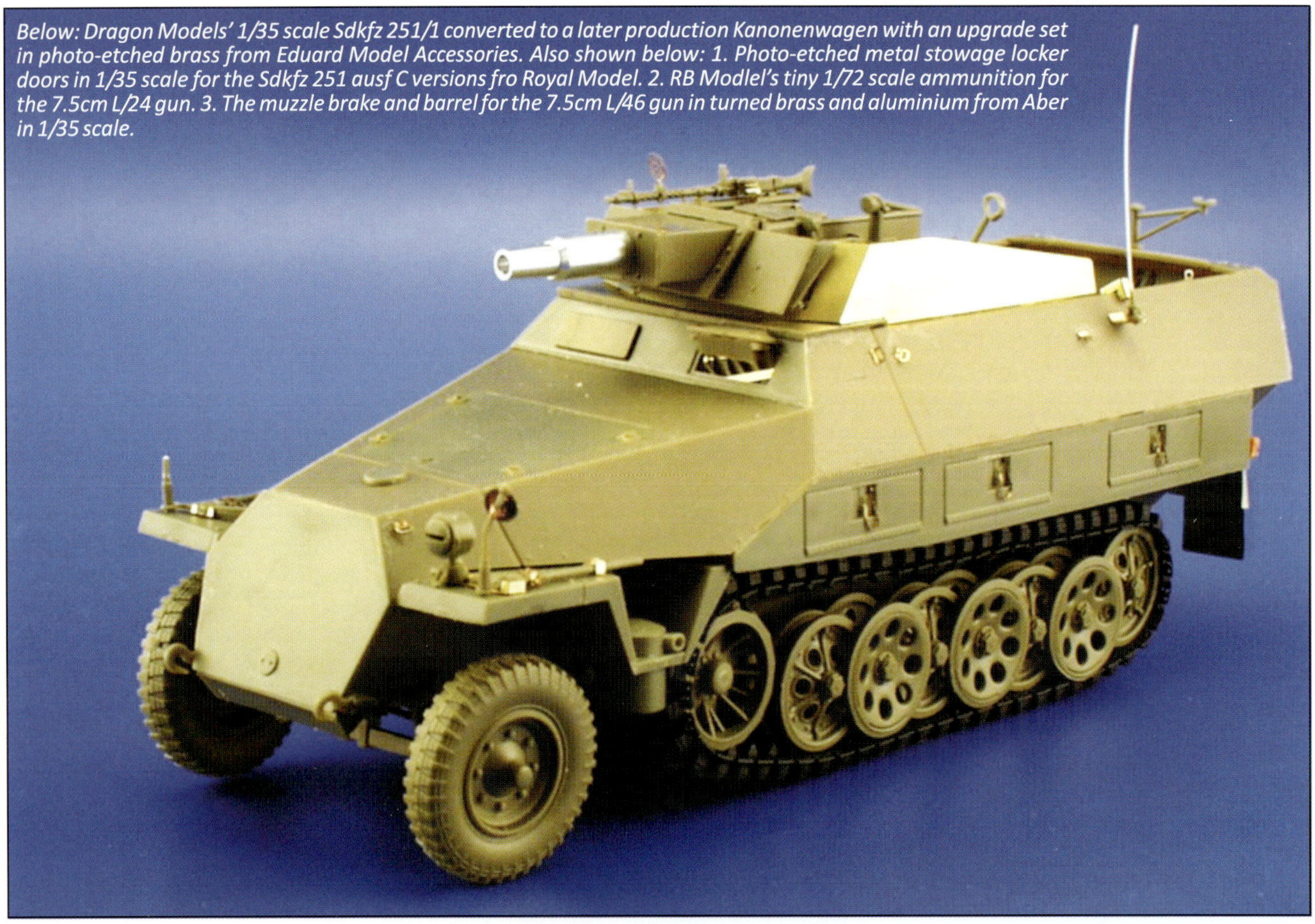

Below: Dragon Models' 1/35 scale Sdkfz 251/1 converted to a later production Kanonenwagen with an upgrade set in photo-etched brass from Eduard Model Accessories. Also shown below: 1. Photo-etched metal stowage locker doors in 1/35 scale for the Sdkfz 251 ausf C versions fro Royal Model. 2. RB Modlel's tiny 1/72 scale ammunition for the 7.5cm L/24 gun. 3. The muzzle brake and barrel for the 7.5cm L/46 gun in turned brass and aluminium from Aber in 1/35 scale.

1

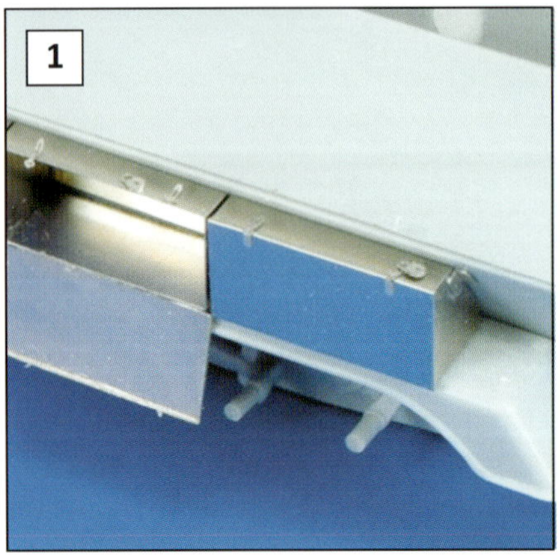

2

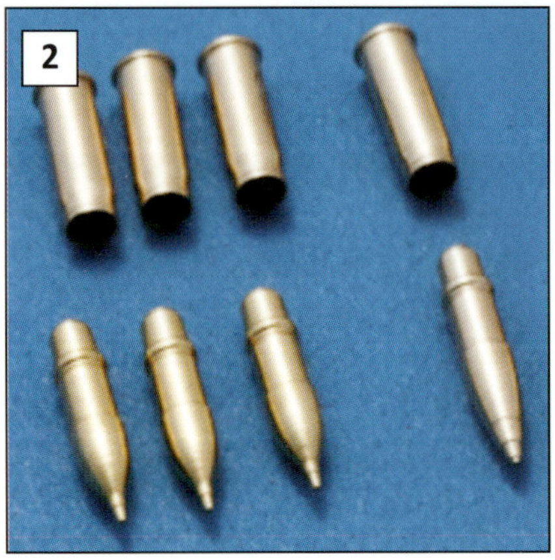

3

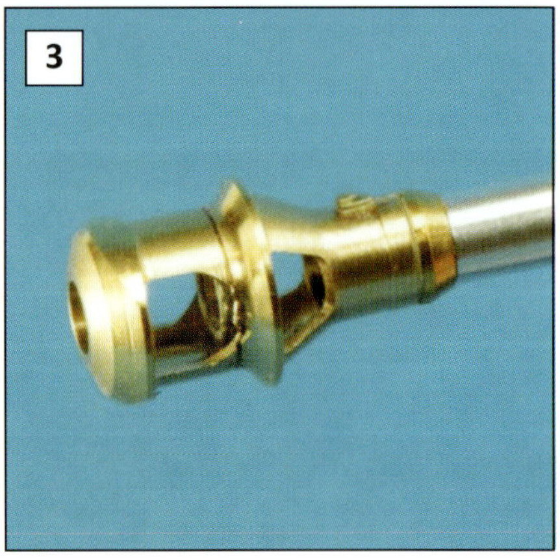

Royal Model. This Italian company has been producing high quality accessories and aftermarket parts since the early 1990s. The current catalogue includes a number of upgrade sets in 1/35 scale for the Sdkfz 251 series of halftracks. While these could be of use they are not specifically made fro the Sdkfz 251/9 and 251/22.

Voyager Model. Shanghai-based company Voyager Model have been manufacturing comprehensive upgrade sets for scale models since 2003 with the release of their first set for 1/35 scale armour. The company also produces turned aluminium barrels with muzzle brakes in 1/35 scale and accessories in resin.

Aber. This Polish company manufactures an extensive range of upgrade sets in photo-etched brass, milled aluminium and brass, stainless steel and even wood. For some time now Tamiya has included Aber products with a number of their models.

A full list of the accessories made specifically for the Sdkfz 251 series of halftracks in all scales would too large to reproduce in this book and the reader can find the company's contact details on page 64.

RB Model. Based in Poland, this company produces a range of milled aluminium and brass gun barrels and resin accessories for 1/35 scale armour models.

Eduard. This Czech company manufactures high-quality accessories in photo-etched brass and resin and was one of the earliest firms to offer upgrade sets in the smaller scales. In addition to the upgrade sets Eduard offers paint masks and markings.

The official organisation of a Panzergrenadier division in mid-1944 authorised two motorised infantry regiments and each was supposed to contain a Gepanzert or armoured infantry company carried in armoured halftracks. In reality almost all the infantry of the army's Panzergrenadier battalions, including those attached to Panzer divisions, relied on trucks and lorries for their transport.

A number of Panzergrenadier divisions did, however, receive the Sdkfz 251/9 halftracks that were authorised under KstN 1114c and in 1945 some were allocated a number of Sdkfz 251/22 vehicles. I should mention that this list may not be complete and I have only included confirmed numbers.

3.Panzergrenadier-Division. The division served in Italy until August 1944 when it was transferred to France and attached to XXXXVII.Armeekorps fighting in the Moselle region. The division was withdrawn from the front and sent to Germany to be rebuilt in November and December 1944 and at that time received six Sdkfz 251/9 7.5cm-armed halftracks. In Panzergrenadier divisions at this time these vehicles were normally allocated to the heavy platoons of Panzergrenadier companies, with two halftracks authorised for each platoon.

The division took part in the Ardennes Offensive and was eventually destroyed in the Ruhr Pocket in April 1945.

10.Panzergrenadier-Division. The division served in the East, suffering heavy losses, until August 1944 when it was withdrawn from the front to the area around Kishinev, Chisinau in modern-day Moldava, and after being partially rebuilt returned to the fighting where it operated as two separate Kampfgruppen for some time.

In January 1945 the division was attached to XXXXVIII.Armeekorps and fought in the battles near Radom south of Warsaw where it was largely destroyed. The survivors were withdrawn to Görlitz in Germany and used to rebuild the division and at this time a number of Sdkfz 251/22 halftracks were received.

Partly rebuilt, the division returned to the front in March 1945 and took part in the fighting in Upper Silesia eventually surrendering to the Soviets at Olomouc in eastern Czechoslovakia.

18.Panzergrenadier-Division. Almost completely destroyed in the summer 1944 battles on the Eastern Front, the remnants of this formation were used to form Panzer-Brigade 105.

When the brigade was disbanded the surviving elements were used to rebuild 18.Panzergrenadier-Division. In January 1945 the division took part in the defence of East Prussia and after fighting around Heilingenbeil in early March the division was once again dissolved. On 30 March 1945 the division was reorganised according to the Kampfgruppe Panzer

At left: An Sdkfz 251/22 of 18.Panzergrenadier-Division disabled and abandoned outside Berlin in April 1945. Note the late production engine access hatches, the lower of which is partly open. Below: An Sdkfz 251/22 halftrack photographed just after the end of the war. This vehicle has the later versions on the rear machine-gun support arm and the Heer number plate. The halftrack in the background is an 250/9 Neu and behind that an Sdkfz 234/3 armoured car.

Division 45 Grundgliederung issued on 25 March 1945, one of very few formations to be rebuilt using this structure. The new formation contained the partly-formed Panzer-Division Schlesien which had a number of Sdkfz 251/22 halftracks on hand.

A halftrack of the division is shown on page 26 of the Camouflage & Markings section.

20.Panzergrenadier-Division. During the summer of 1944 the division was attached to 1.Panzerarmee and fought in the counterattack at Lemberg and in the defence of Kamionka Strumilowa where it was encircled for five days. From August the division was engaged in the fighting along the Vistula and the defensive battles in front of Warsaw.

In December 1944 the division was withdrawn from the front and transferred to the area around Kielce, in southern Poland, reporting that a single Sdkfz 251/9 halftrack was on hand.

In early 1945 the division was briefly subordinated to the recently-created Panzerkorps-Grossdeutschland and then fought on the Oder Front south-west of Küstrin. Most of the division was destroyed in the Halbe Pocket.

25.Panzergrenadier-Division. The division was almost completely destroyed in the summer 1944 battles on the Eastern Front and the survivors were incorporated into Panzer-Brigade 107.

The brigade was dissolved in November with the remnants used to build a new 25.Panzergrenadier-Division. The division reported that eight Sdkfz 251/9 halftracks were on hand at the end of December 1944 and these may have come from Panzer-Brigade 107. During January 1945 the division took part in the defensive battles in western Germany and at that time a number of Sdkfz 251/22 halftracks were received.

The division was transferred to the Eastern Front in the following month and fought on the Oder Front.

Panzergrenadier-Division(GD) Grossdeutschland. This formation, as many readers will know, was in actuality an over-strength Panzer division and in late September 1944 it was combined with Panzergrenadier-Division Brandenburg to form Panzerkorps-GD.

Both the division's Panzergrenadier-Regiment GD and Panzerfüsilier-Regiment GD contained an armoured infantry battalion and both operated numbers of Sdkfz 251/9 halftracks (1).

In addition, the division's Aufklärungs battalion had a full allocation of 7.5cm-armed halftracks. On 30 December 1944 the division reported that sixteen Sdkfz 251/9 halftracks were on hand and it is likely that six of these were with the reconnaissance battalion, two less than its authorised total. At the same time the division reported a number of Sdkfz 251/22 halftracks on hand and this was probably the first formation to receive the type.

Just two weeks later, on 12 January 1945, the units of Panzerkorps-GD were ordered to convert all their existing Sdkfz 251/9 halftracks to the Sdkfz 251/22 but is not known if enough conversion kits were ever received.

Vehicles of this division are shown on pagcs 21 and 24 of the Camouflage & Markings section.

Panzergrenadier-Division Brandenburg. Formed in September 1944, the division was made up of two infantry units, referred to as Panzer-Jäger-Regiment 1 and Panzer-Jäger-Regiment 2 (2), an artillery element and an Aufklärungs battalion. The two Jäger regiments were reduced from three to two battalions when the division was officially attached to Panzerkorps-GD in December 1944.

Photographic evidence confirms that the division operated a number of Sdkfz 251/9 halftracks and it is likely that these were allocated to one or both of the Jäger regiments as the reconnaissance battalion seems to have been equipped with versions of the Sdkfz 234 armoured car.

On 12 January 1945 the divisions of Panzerkorps-GD were ordered to upgrade all their Sdkfz 251/9 halftracks by fitting the 7.5cm L/46 gun but it seems that no conversion kits were issued before the war ended.

A halftrack of this division is shown on page 22 of the Camouflage & Markings section.

Panzergrenadier-Division Kurmark. Formed in late January 1945, the division only ever reached the strength of a brigade.

Commanded by Generalmajor Willy Langkeit, who had served with the GD division, Panzergrenadier-Division Kurmark took part in the battles around Frankfurt an der Oder in February and in late march the division was withdrawn from the front and rebuilt.

In mid-April the division was attached to 9.Armee and although most of its personnel were cut off and surrounded in the Halbe Pocket south of Berlin, a number escaped to the West.

The second company of the division's Panzer-Aufklärungs-Abteilung 151 was probably equipped with two Sdkfz 251/9 halftracks. This unit is sometimes referred to as Panzer-Aufklärungs-Schwadron (gemischte) Kurmark.

Notes

1. Both regiments were reduced to two battalions at this time and a Korps-Fusilier Regiment GD was raised. This unit was often confused with Panzerfüsilier-Regiment GD and in March 1945 it was renamed Panzermusketier-Regiment GD.
2. Not to be confused with Panzerjäger or anti-tank units. The use of the word Jäger here refers to light infantry.

At right: An Sdkfz 251/9 halftrack of 2.SS-Panzer-Division. The tactical symbol is not visible but this vehicle is almost certainly from one of the Panzergrenadier companies. Above: A detail of the hull front plate of one of the Sdkfz 251/1 halftracks of SS-Panzergrenadier-Regiment 4 showing the tactical symbol identifying an armoured infantry company and the division's unit insignia.

The SS-Führungshauptamt did issue a number of KstN during the war that related specifically to units of the Waffen-SS, such as the series issued in October 1944 authorising the creation of SS-Jäger-Bataillonen, but for the most part the Panzer and Panzergrenadier divisions were organised using the standard OKW orders.

The histories of these formations are widely available and well known and I have restricted the narrative here to aspects relating to the allocation of Sdkfz 251/9 and 251/22 armoured halftracks. The titles of several of these formations changed throughout the war and those employed here are correct for the 1944-1945 period.

1.SS-Panzer-Division Leibstandarte SS Adolf Hitler (LSSAH). The premier division of the Waffen-SS, this formation took part in the Normandy battles during the summer of 1944. Although a total of fourteen Sdkfz 251/9 halftracks were recorded as in transit to the division on 1 July 1944, neither III.Bataillon, SS-Panzergrenadier-Regiment 2, the division's armoured infantry battalion, nor SS-Panzer-Aufklärungs-Abteilung 1 received any before the end of the campaign.

The division was partly rebuilt and took part in the Ardennes Offensive with its various elements organised into a series of battle groups to better negotiate the close terrain. Attached to Kampfgruppe Peiper, III.Bataillon, SS-Panzergrenadier-Regiment 2 started the battle with nineteen Sdkfz 251/9 halftracks on hand. This battalion was reinforced in a similar manner to I.Bataillon, Panzergrenadier-Regiment 14 of 5.Panzer-Division, as mentioned above, and again this was probably a temporary measure. Altogether the regiment lost forty-nine of its armoured halftracks in the battle and although I have not been able to determine the exact types, a report dated 30 December 1944 gives a total of only six 7.5cm-armed vehicles for the whole division.

In January 1945 the division was withdrawn from the front line and in the following month sent to Hungary, taking part in Operation Frühlingserwachen, or Spring Awakening, the last major German offensive of the war. Withdrawing towards the west the division was engaged in the unsuccessful defence of Vienna and eventually reached Steyr, south of Linz in Austria, where the survivors surrendered to units of the US Army on 8 May 1945. A vehicle of this formation is shown on page 23 of the Camouflage & Markings section.

2.SS-Panzer-Division Das Reich. The division was stationed in southern France when the Allied invasion began on 6 June 1944 and took some time to arrive at the front.

On 1 July 1944, III.Abteilung, SS-Panzergrenadier-Regiment 4 reported that twelve Sdkfz 251/9 halftracks were on hand. The division's SS-Panzer-Aufklärungs-Abteilung 2 was supposed to be re-equipped with Sdkfz 233 armoured cars in late 1943 but it appears that at least two Sdkfz 251/9 halftracks were retained by the battalion's 5.Kompanie and one of these vehicles is shown on page 18 of the Camouflage & Markings section of this book.

In early June 1944 the battalion's fifth and fourth companies were amalgamated to form a 4.Kompanie. Almost all the division's equipment was lost in the Normandy battles and the subsequent retreat into Belgium and in October 1944 the survivors were transferred to Germany where 2.SS-Panzer-Divison was rebuilt.

The division took part in the Ardennes Offensive and SS-Panzergrenadier-Regiment 4 reported that twelve Sdkfz 251/9 halftracks were available on 16 December 1944 when the assault began. At the end of the same month the division had five halftracks on hand and ten in transit and these vehicles were probably with the division when it was transferred to the East in early February 1945.

3.SS-Panzer-Division Totenkopf. Formed in late 1939, the division served exclusively on the Eastern Front. During the summer of 1944 the division was vastly understrength and a report compiled by the divisional headquarters and dated 1 June gives the authorised number of Sonderpanzerwagen (SPW), or armoured halftracks, as 367 when the actual number available for immediate deployment was twenty-three. At this time the division's SS-Panzer-Aufklärungs-Abteilung 3 was supposed to contain a number of Sdkfz 250/9 halftracks but none of these was actually available (1).

Allocations of new vehicles must have been made in the autumn and winter as this formation was the strongest armoured division on the Eastern Front by December 1944. At the end of the month the division reported that ten Sdkfz 251/9 halftracks were on hand and these were almost certainly allocated to I.Bataillon, SS-Panzergrenadier-Regiment 5.

5.SS-Panzer-Division Wiking. During the fighting in central Poland in 1944, the division's SS-Panzergrenadier-Regiment 9 Germania was equipped with a number of Sdkfz 251/9 halftracks and this is explained in the chart on the following page. At the end of December 1944, when the division was sent to Hungary, the number of Sdkfz 251/9s on hand had risen to fourteen. A halftrack of SS-Panzergrenadier-Regiment 9 is shown on page 17 of the Camouflage & Markings section.

9.SS-Panzer-Division Hohenstaufen. In June 1944 the division's SS-Panzer-Aufklärungs-Abteilung 9 reported that four Sdkfz 251/9 halftracks were on hand with 5.Kompanie. At the same time III.Bataillon, SS-Panzergrenadier-Regiment 20, the division's armoured infantry battalion, had eight 7.5cm-armed halftracks distributed between its four companies. The division was engaged in the defensive battles in the Netherlands in September and reinforced for the Ardennes Offensive which commenced on 16 December 1944. Sources differ on the number of vehicles available but it is likely that the heavy company of III.Bataillon, SS-Panzergrenadier-Regiment 20 had five Sdkfz 251/9 halftracks while a further five, or possibly six, were allocated to the other three companies. SS-Panzer-Aufklärungs-Abteilung 9 had four vehicles on hand with its fifth company.

10.SS-Panzer-Division Frundsberg. During the Normandy battles the division's SS-Panzer-Aufklärungs-Abteilung 10 and the first battalion of SS-Panzergrenadier-Regiment 21 were equipped with Sdkfz 251/9 halftracks but I have been unable to confirm the exact numbers. On 16 December 1944, the day the Ardennes Offensive began, SS-Panzergrenadier-Regiment 21 reported that just one 7.5cm-armed halftrack was on hand but this number had risen to twelve by the end of the month. A halftrack vehicle of the division is shown on page 20 of the Camouflage & Markings section.

Notes

1. This was a version of the light Sdkfz 250 halftrack armed with a 2cm gun in a revolving turret. These vehicles were specifically designed to equip reconnaissance units.

Photographed in late July 1944 near the village of Wilanowo, south of Bialystok on the Bug river, this Sdkfz 251/9 halftrack is one of the 7.5cm-armed vehicles allocated to III.Bataillon, SS-Panzergrenadier-Regiment 9 Germania. Other halftracks of this battalion are shown in the Camouflage & Markings section on page 17.

SS-PANZERGRENADIER-REGIMENT 9 GERMANIA

During the summer months of 1944, 5.SS-Panzer-Division was one of the strongest armoured units fighting on the Eastern Front. Almost completely rebuilt after the winter battles for the Cherkassy Pocket, the division's III.Bataillon, SS-Panzergrenadier-Regiment 9 was completely armoured and its actual strength is shown here as a typical Gepanzert battalion of the time. Note that I have not included the battalion's Versorgungs-Kompanie or the many cars and motorcycles that were also on hand. The organisation depicted here should be compared with the official establishments shown in the section dealing with Kriegsstärkenachweisungen (KstN) which begins on page 56.

The battalion's 12.Kompanie, the heavy company, was unusual in that it was composed of parts of other units referred to in German as Teileinheiten and abbreviated as TE. The KstN were initially issued as ohne nummer, or unnumbered, and were almost certainly drafts which were later replaced. The company headquarters was organised according to Führer einer schweren Kompanie a (gepanzert) (frei) KstN ohne nummer von 1.04.1944 and this was probably superseded at some later date by (TE) KstN 1121 a (gp) (fG) von 1.4.1944. The company had previously been organised using (TE) Führer einer schweren Kompanie a (gepanzert) KstN 1121a (gp) von 1.11.1943. The first platoon, or 12-cm Granatwerfer-Zug (gepanzert) KstN ohne nummer von 1.04.1944, had not received its allocation of 12cm mortars or halftracks when the battalion was sent into combat. The second platoon was built using schwerer Panzerjäger-Zug (zu 3 Geschützen) (gepanzert) KstN 1145 (gp) von 1.11.1943. This establishment was in fact the motorised version of KstN 1145, equipped with 7.5cm Pak 40 anti-tank guns, but towed by Sdkfz 251 halftracks in place of trucks. The third platoon was unusual in being equipped with Sdkfz 251/7 Pionierwagen and may have replaced (Teileinheit) Geschutz-Zug a (zu 2 leichten Jnfanterie-Geschutzen) (gepanzert) KstN 1123a (gp) von 1.11.1943. The fourth platoon was organised according to Schwerer Kanonen-Zug (7.5 cm) (gepanzert) (freie Gliederung) ohne nummer von 1.04.1944. Another draft order, this KstN was replaced by (Teileinheit) schwere Kanonen-Zug (gepanzert) (frei Gliederung) KstN 1125 (gp)(fG) von 1.4.1944 described in detail on page 57. The platoon had originally been built using (Teileinheit) schwerer Kanonen-Zug (7.5 cm) KstN 1125 (gp) von 1.11.1943.

An Sdkfz 251/9 halftrack of I.Abteilung, SS-Panzergrenadier-Regiment 21 photographed shortly after a Kampfgruppe made up from the remnants of 10.SS-Panzer-Division attempted to fight their way towards the Belgian border after being encircled near Bapaume in early September 1944. Inset is an enlarged view of the division's unit insignia. This vehicle is also shown in the Camouflage & Markings section on page 20.

11.SS-Freiwilligen-Panzergrenadier-Division Nordland. The division was formed in early 1943 with the intention that it would include all the Scandinavian and Dutch volunteer formations which had been raised up to that time.

The division served in Croatia and was sent to the Leningrad Front in November 1943 and suffered heavy casualties during the winter battles along the Luga and Narva Rivers. In 1944 the Nordland division was engaged in the defensive battles on the Tannenburg Line and later in the fighting in Kurland in what is today western Latvia. By early 1945 the division had been pushed back to Pomerania where it remained until March when it was transferred to the Oder Front and by April the survivors were fighting in Berlin.

From its formation the division's SS-Panzer-Aufklärungs Abteilung 11 was equipped with a number of Sdkfz 251/9 halftracks, reporting that five were on hand with 5.Kompanie at the beginning of 1944. Three of these were said to have been lost on 26 January in an action at Gubanitsy, about 50 kilometres south-west of modern-day St Petersburg, in the fighting for the Narva Bridgehead. One of the company's vehicle commanders, Unterscharführer Casper Antoine Sporck, is credited with destroying nine T-34 tanks during this battle. In late August 1944 at Dorpat, modern-day Tartu in Estonia, Sporck's halftrack almost single-handedly saved the German defence, a feat for which he was awarded the Ritterkreuz, one of only three Dutch volunteers to receive the medal. By the end of the year 5.Kompanie reported that four Sdkfz 251/9 halftracks were on hand although one was in repair.

12.SS-Panzer-Division Hitlerjugend. Formed in mid-1943 the division served in Normandy, the Ardennes and on the Eastern Front.

At the beginning of the Allied invasion in June 1944, III. Bataillon of the division's SS-Panzergrenadier-Regiment 26 had a number of Sdkfz 251/9 halftracks but I have not been able to discover the exact total. Also in June, 5.Kompanie, SS-Panzer-Aufklärungs-Abteilung 12 reported six vehicles on hand but by the end of the month the battalion's fourth company had absorbed the remnants of 5.Kompanie and it seems that just one Sdkfz 251/9 survived the Normandy battles.

In November 1944 the division was rebuilt, with the Aufklärungs battalion reporting that three halftracks were on hand but there is some debate as to how they were allocated to the companies. On 18 December 1944, III.Bataillon, SS-Panzergrenadier-Regiment 26 reported that a total of fourteen Sdkfz 251/9 halftracks were available, distributed among its four companies. On 30 December 1944, when the Ardennes Offensive was almost concluded, the division reported that a total of seventeen 7.5cm-armed halftracks were available. A report of 1 March 1945 states that 12.Kompanie, SS-Panzergrenadier-Regiment 26 had received a number of Sdkfz 251/22 vehicles and these part of a shipment of twenty-five halftracks that the division received in February.

A halftrack vehicle of the unit division formation is shown on page 19 of the Camouflage & Markings section.

Infanterie-Division Ulrich von Hutten. Formed at the end of March 1945, the division went into action within weeks near Bitterfeld, north of Leipzig, against units of the US Army before moving towards Berlin where it was eventually trapped in the Halbe Pocket south of the city in April. Official Soviet photographs purport to show disabled Sdkfz 251/9 halftracks of the division abandoned near Berlin but I would be reluctant to accept this identification at face value but the division did have a varied assortment of vehicle types.

Sturm-Panzer-Abteilung 217. The battalion was formed in June 1944 and equipped with the Sturmpanzer IV self-propelled 15cm howitzer. Made up of three companies, the battalion fought in Normandy and the Ardennes Offensive and was destroyed in the Ruhr Pocket battles in March 1945. The battalion reported that two Sdkfz 251/9 halftracks were on hand on 30 December 1944.

Heeres-Pionier-Sturmbrigade 627 (motorisiert). Formed in October 1944 by absorbing most of the semi-independent Pionier units that had been fighting in the Warsaw area, the brigade was made up of two battalions until April 1945 when a third battalion was raised from Heeres-Sturm-Pionier-Bataillon (mot) 500. A report of 30 December states that four Sdkfz 251/9 halftracks were on hand at that time. Commanded by Oberstleutnant Karl Herzog, the brigade took part in the battles in Upper Silesia, most notably around Mackheim, modern-day Maków Mazowiecki, in mid-January 1945 where Oberstleutnant Herzog personally led the halftracks in a successful counterattack to secure the Baranowo-Praschnitz road. The last remnants of the brigade were cut off at Peyse, today a suburb of Svetly on the Samland Peninsula, and surrendered to Red Army units on 17 April 1945.

Fallschirm-Panzer-Division 1 Hermann Göring. After serving in Italy the division was transferred to the Eastern Front in July 1944 and took part in the fighting around Warsaw with 5.SS-Panzer-Division Wiking and 19.Panzer-Division. In October the division was moved to East Prussia and at the end of 1944 reported that six Sdkfz 251/9 halftracks were on hand. These were almost certainly the vehicles that arrived in October and were allocated to Fallschirm-Panzer-Aufklärungs-Abteilung 1 Hermann Göring. There is no evidence to support the claim that this division operated a number of Sdkfz 251/22 halftracks.

Panzer-Aufklärungs-Abteilung Speer. Raised from members of the Reichsarbeitdienst (RAD), the battalion was made up of a motorcycle company, equipped with Sdkfz 2 Kettenkrad and an armoured car company with a number of Sdkfz 251/9 halftracks. The battalion was attached to the curiously-named Kampfgruppe 1001 Nacht which supported 25.Panzergrenadier-Divivison in the defence of Wriezen, south-east of Eberswalde, on the Oder Front in mid-April. Most of Kampfgruppe 1001 was trapped in the Halbe Pocket and destroyed.

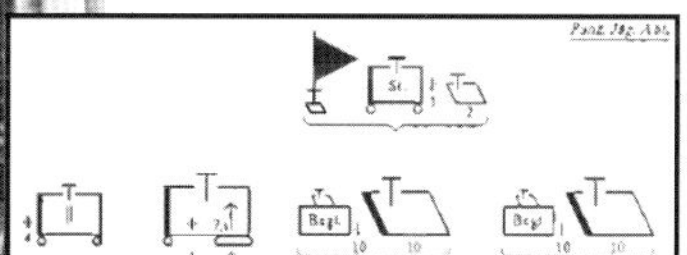

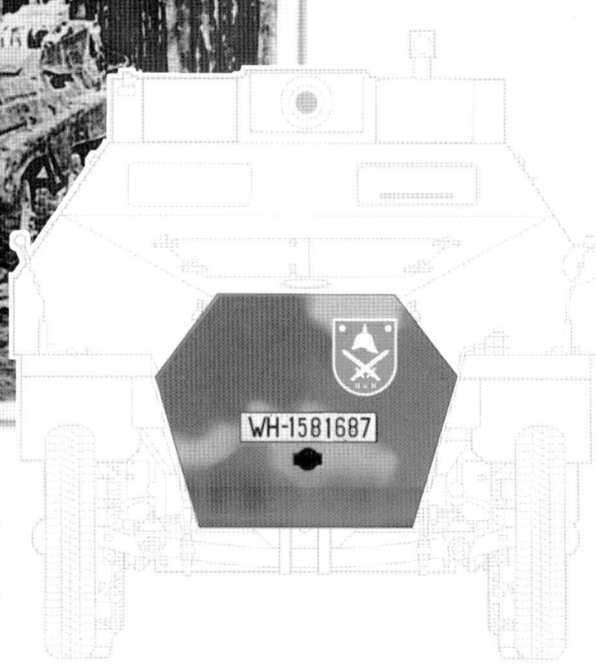

Above: Although this photograph is of rather poor quality it is interesting in that it may show vehicles of Infanterie-Division Ulrich von Hutten, a very short-lived formation. At right is a possible reconstruction of the Sdkfz 251/9 halftrack and above that is a detail from the actual Grundgliederung for the Panzer-Division 45 issued on 25 March 1945 which shows the compostion of the division's Panzerjager-Abteilung. Note the symbol second from left which denotes a company equipped with 7.5cm-armed halftracks.

KRIEGSSTÄRKENACHWEISUNG (KSTN) 1114C

All formations of the German Army were organised according to detailed tables of establishment referred to as Kriegstärkenachweisung (KstN) which showed the official composition of a unit, listing the exact number of personnel and type of equipment from vehicles to small arms. The organisation depicted below was based on KstN 1114 which had first been introduced in 1937. It was used to build motorised infantry companies or reconnaissance squadrons, as they were then known, and Sdkfz 251/9 halftracks were first authorised in KstN 1114c issued on 1 November 1943. Although this organisation remained largely unchanged until the end of the war, it was issued in at least fourteen amended versions and there were actually two April 1944 variants with the second incorporating the so-called freie Gliederung aspects. The concept of freie Gliederung, or free organisation, which was introduced in early 1944, saw many support elements, such as transport and supply units, transferred from company to battalion or regimental command. The allocation of Sdkfz 251/17 halftracks to the platoon commanders was overly optimistic as the first of these vehicles did not leave the factories until November 1944 and most units were issued with Sdkfz 251/10 3.7cm-armed halftracks instead. The table shown here shows the organisation authorised in late April 1944 and, below that, the company's place with the Aufklärungs battalion of a Panzer division.

Panzergrenadier-Kompanie ausf c (gepanzert) (frei Gliederung)
oder
Panzer-Aufklärungs-Kompanie (gepanzert) (frei Gliederung)
KstN 1114c (gp)(fG) von 1.4.1944

Gruppe Führer und Kompanietrupp

2 x Kfz 2 VW Car

3 x Sdkfz 2 Kettenkrad

2 x Sdkfz 251/3

1.Zug

1 x Sdkfz 251/17

2.Zug

As for 1.Zug

3.Zug

As for 1.Zug

1.Gruppe

3 x Sdkfz 251/1

2.Gruppe

3 x Sdkfz 251/1

3.Gruppe

3 x Sdkfz 251/1

4.(Schwere) Zug

1 x Sdkfz 2 Kettenkrad

1 x Sdkfz 251/17

Granatwerfer Gruppe

2 x Sdkfz 251/2

Kanonen Gruppe

2 x Sdkfz 251/9

Stab und Stabs-Kompanie
Panzer-Aufklärungs-Abteling (gepanzert) (frei Gliederung)
KstN 1109 (gp) (fG) von 1.4.1944

Versorgungs-Kompanie Panzer-Aufklärungs-Abteilung
(gepanzert) (frei Gliederung)
KstN 1180 (gp)(fG) von 1.4.1944

Panzer-Spähwagen-Kompanie ausf c
(frei Gliederung)
KstN 1162c (fG) von 1.4.1944

1.Zug *2.Zug* *3.Zug* *4.Zug*

Leichte Panzer-Aufklärungs-Kompanie
(gepanzert) (frei Gliederung)
KstN 1113 (gp)(fG) von 1.4.1944

1.Zug *2.Zug* *3.Zug* *4.Zug*

Panzergrenadier-Kompanie
ausf c (gepanzert) (frei Gliederung)
oder
Panzer-Aufklärungs-Kompanie
(gepanzert) (frei Gliederung)
KstN 1114 (gp)(fG) von 1.4.1944

1.Zug *2.Zug* *3.Zug* *4.Zug*

1. (Teileinheit) Führer schwere Kompanie ausf a (gepanzert) (frei Gliederung)
(TE) KstN 1121 ausf a (gp)(fG) von 1.4.1944

2. (Teileinheit) schwere Kanonen-Zug (gepanzert) (frei Gliederung)
(TE) KstN 1125 (gp)(fG) von 1.4.1944

3.(Teileinheit) 8cm Granatwerfer-Zug (6 Granatenwerfer) (gepanzert) (frei Gliederung)
(TE) KstN 1126 ausf a (gp)(fG) von 1.4.1944

4. (Teileinheit) Pionier-Zug ausf a (gepanzert) (frei Gliederung)
(TE) KstN 1124 ausf a (gp)(fG) von 1.4.1944

The battalion's fourth company, shown at left, was made up of parts, or Teileinheiten, of four different establishments. The full version of KstN 1125 is detailed on the following page.

KRIEGSSTÄRKENACHWEISUNG (KSTN) 1125

The Kanonen Züge, or gun platoons, of heavy companies were organised using KstN 1125 which was first introduced in late 1942. These units were intended to give extra firepower to infantry and reconnaissance formations which often found themselves far in advance of the division's artillery support. This table of establishment went through relatively few changes during the war with the most important amendments involving the introduction of the freie Gliederung principles where ancillary elements were concentrated in higher commands. Note that the complete company was made up of Teileinheiten, or parts, of other organisations.

(Teileinheit) Führer schwere Kompanie ausf a (gepanzert) (frei Gliederung) (TE) KstN 1121 ausf a (gp)(fG) von 1.4.1944

The company command also contained 3 x Sdkfz 2 Kettenkrad and 2 x Kfz 1 Light Cars

Sdkfz 251/3

Sdkfz 251/11

(Teileinheit) schwere Kanonen-Zug (gepanzert) (frei Gliederung) (TE) KstN 1125 (gp)(fG) von 1.4.1944

Zug Trupp

Sdkfz 251/3

Sdkfz 251/1

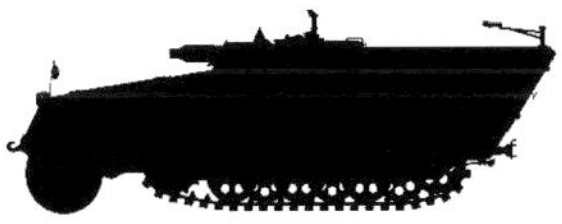

Sdkfz 251/9

Sdkfz 251/9

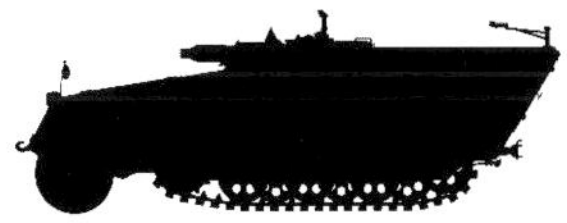

Sdkfz 251/9

Sdkfz 251/9

Sdkfz 251/9

Sdkfz 251/9

Although all these organisations were amended in November 1944, the most significant change was made to KstN 1125 which lost the Sdkfz 251/1 of the platoon headquarters and was authorised six Sdkfz 251/22 halftracks in place of the vehicles shown here.

(Teileinheit) 8cm Granatwerfer-Zug (6 Granatenwerfer) (gepanzert) (frei Gliederung) (TE) KstN 1126 a (gp)(fG) von 1.4.1944

2 x Sdkfz 251/1, 6 X Sdkfz 251/2 8cm Mortar

(Teileinheit) Pionier-Zug ausf a (gepanzert) (frei Gliederung) (TE) KstN 1124 a (gp)(fG) von 1.4.1944

1 x Sdkfz 251/2 8cm Mortar, 7 x Sdkfz 251/7 Pionierwagen

In some units the Granatwerfer Zug was replaced by (Teileinheit) 12cm Granatwerfer-Zug (6 Granatenwerfer) (gepanzert) (frei Gliederung) (TE) KstN 1126 (gp) von 1.4.1944 with six towed 12cm heavy mortars.

Stab Panzergrenadier-Bataillon (gepanzert) (frei Gliederung) KstN 1104a (gp) (fG) von 1.4.1944

Versorgungs-Kompanie Panzergrenadier-Bataillon (gepanzert) (frei Gliederung) KstN 1180a (gp)(fG) von 1.4.1944

Panzergrenadier-Kompanie ausf c (gepanzert) (frei Gliederung) oder Panzer-Aufklärungs-Kompanie (gepanzert) (frei Gliederung) KstN 1114 (gp)(fG) von 1.4.1944

1.Zug | 2.Zug | 3.Zug | 4.Zug

Panzergrenadier-Kompanie ausf c (gepanzert) (frei Gliederung) oder Panzer-Aufklärungs-Kompanie (gepanzert) (frei Gliederung) KstN 1114 (gp)(fG) von 1.4.1944

1.Zug | 2.Zug | 3.Zug | 4.Zug

Panzergrenadier-Kompanie ausf c (gepanzert) (frei Gliederung) oder Panzer-Aufklärungs-Kompanie (gepanzert) (frei Gliederung) KstN 1114 (gp)(fG) von 1.4.1944

1.Zug | 2.Zug | 3.Zug | 4.Zug

(Teileinheit) Führer schwere Kompanie ausf a (gepanzert) (frei Gliederung) (TE) KstN 1121 ausf a (gp)(fG) von 1.4.1944

(Teileinheit) schwere Kanonen-Zug (gepanzert) (frei Gliederung) (TE) KstN 1125 (gp)(fG) von 1.4.1944

(Teileinheit) 8cm Granatwerfer-Zug (6 Granatenwerfer) (gepanzert) (frei Gliederung) (TE) KstN 1126 a (gp)(fG) von 1.4.1944

(Teileinheit) Pionier-Zug ausf a (gepanzert) (frei Gliederung) (TE) KstN 1124 a (gp)(fG) von 1.4.1944

KRIEGSSTÄRKENACHWEISUNG (KSTN) 1148

KstN 1148 had been initially promulgated in February 1940 and was used to organise the first self-propelled anti-tank units equipped with the 4.7cm PaK(t) Panzerjäger I. During the next four years it was re-issued in amended formats at least seven times and dealt with the organisation of anti-tank companies including those of the Nashorn and Tiger P units and the Panzerjäger companies of the Schnelle-Abteilungen that were formed for some infantry formations in 1944. However, the Sonderpanzerwagen (SPW) version of 1 April 1945 described here was specifically created to regulate the allocation of Sdkfz 251/22 halftracks to Panzerjäger battalions of Panzer or Panzergrenadier divisions. The organisation of the company is shown in detail, and below that, its place within the battalion.

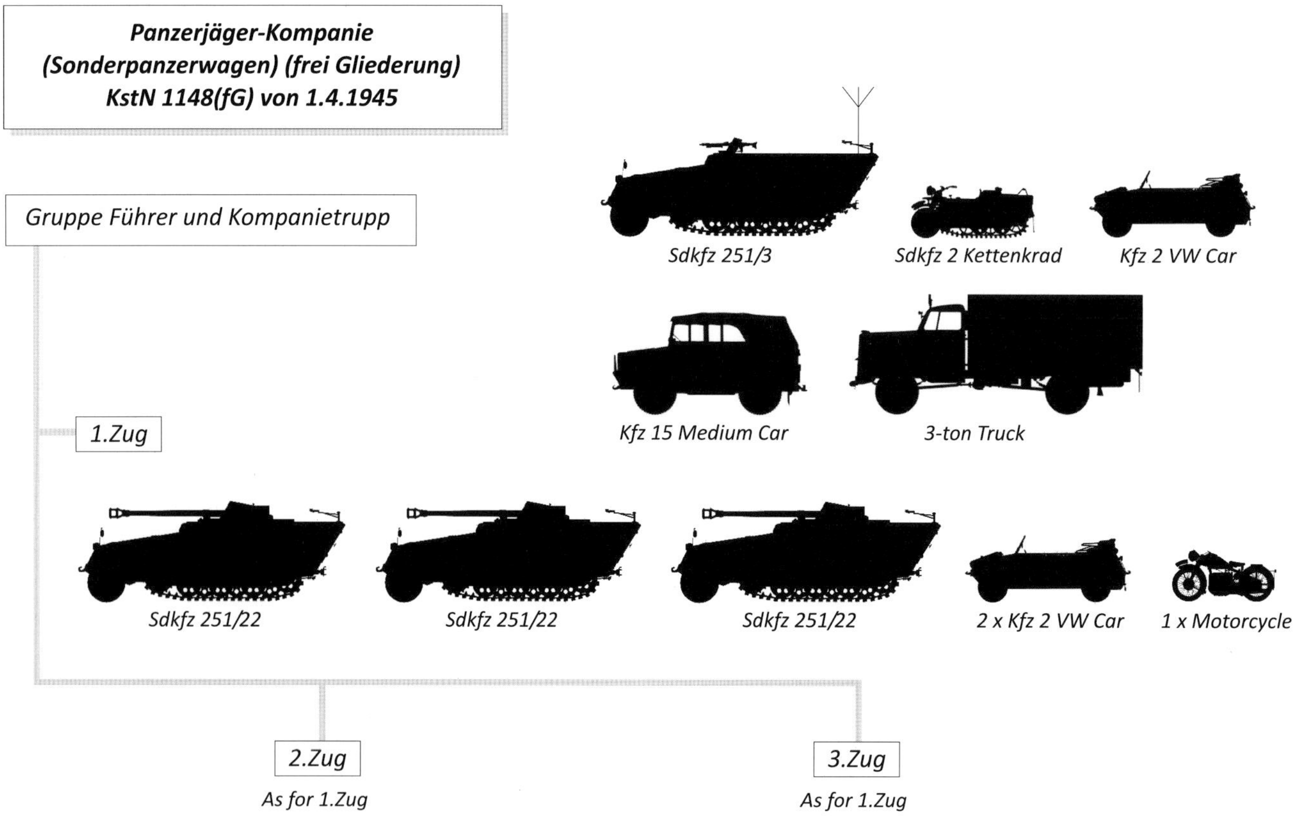

Prior to the promulgation of the 1945 changes the third company of the Panzerjäger battalions of Panzer and Panzergrenadier divisions had been equipped with towed 7.5cm Pak 40 anti-tank guns.

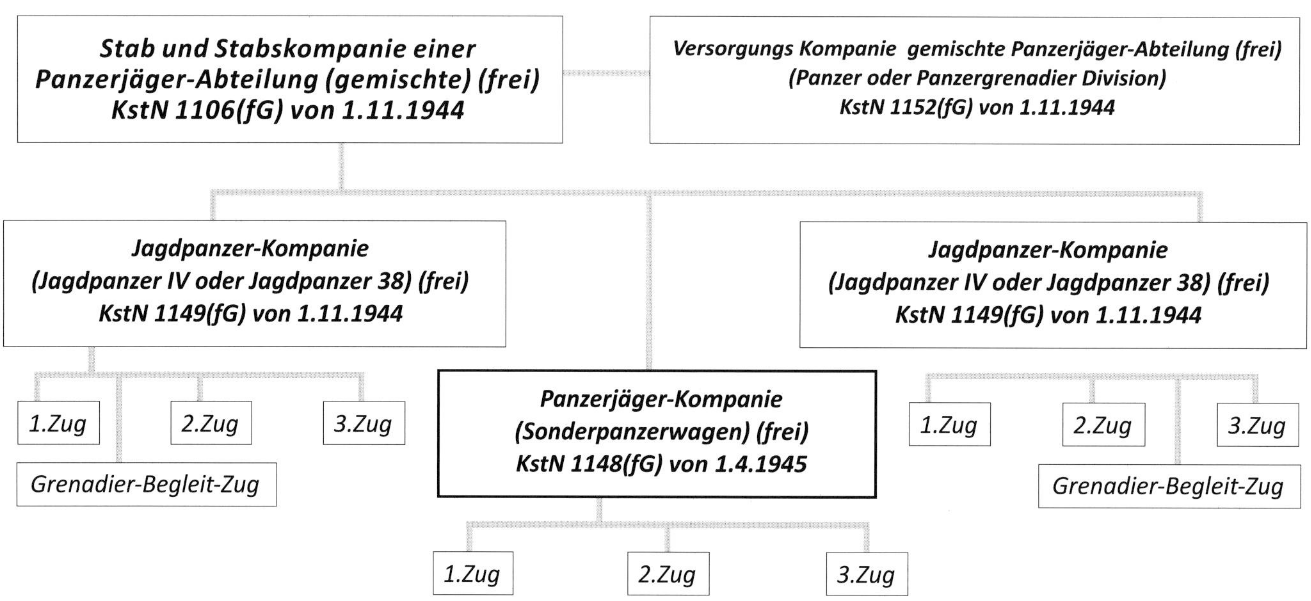

Although the KstN shown here for the Stab und Stabskompanie is given in the OKH order of 25 March 1945, which details the organisation of a 1945 Panzer division, I have been unable to confirm its use or even existence. The known KstN with this number and dated 1 November 1944 refer specifically to the Panzerjäger battalions of Infanterie, Gebirgsjäger and Jäger divisions. It is possible that the Gemischte, or mixed, organisation may be an amended version of KstN 1106(fG) von 1.4.1944 which was in fact marked as provisional when it was first issued.

KstN 1149 (fG) Ausführung A allowed for three platoons of four vehicles each with one for the company commander giving a total of thirteen Jagdpanzer IV or Jagdpanzer 38 while the Ausführung B version had three-vehicle platoons with a total of ten for the company.

Grenadier-Begleit Züge, or Grenadier Escort Platoons, were established by an order issued by the Generalinspekteur der Panzertruppen dated 20 October 1944. These platoons were to be raised for all infantry divisions, including Gebirgs and Jäger formations, equipped with Sturmgeschütz assault guns or Jagdpanzer tank destroyers. They are also included in the detailed Grundgliederung issued for Panzer-Division 1945 and Kampfgruppe Panzer-Division 1945 which was issued on 25 March 1945. The platoon was made up of sixty-one officers and men but in practice their transport was woefully inadequate.

KRIEGSSTÄRKENACHWEISUNG (KSTN) 1162

KstN 1162 was a pre-war establishment that had been used to organise armoured car units since 1936. It had been amended numerous times, most notably with the introduction of the frei Gliederung principles in early 1944. The table shown below, KstN 1162f, was created specifically for the Panzer Division 45 and Kampfgruppe Panzer Division 45 organisations initiated by an order of 25 March 1945 and was an effort to give the reconnaissance battalions a greater tank-killing capability. Given the equipment and vehicles required to fulfil this KstN, at a time when the war was so close to its end, it is doubtful that it was ever implemented, at least in full. As with the previous charts, the organisation of the company is shown in detail, and below that, its place within the battalion.

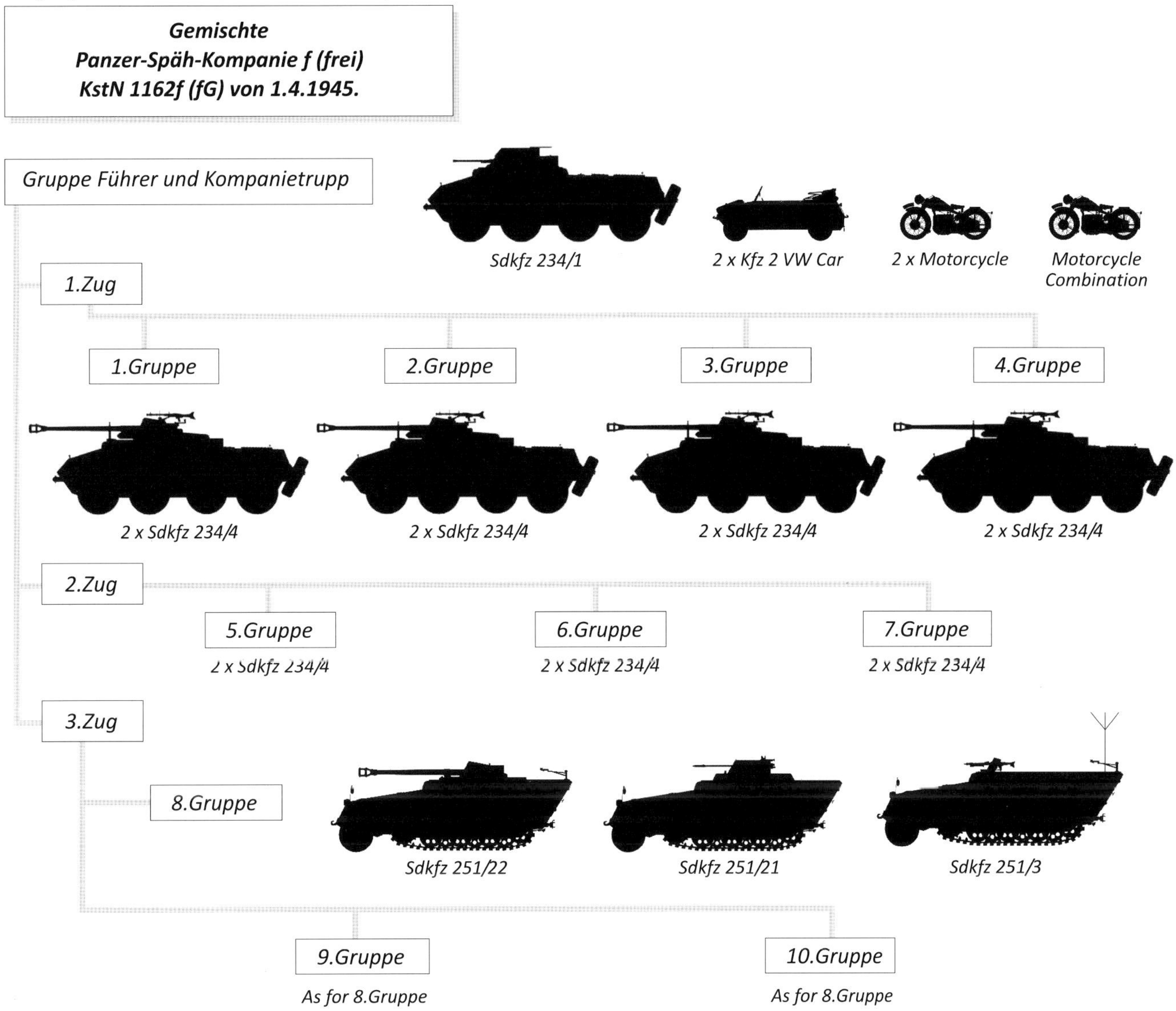

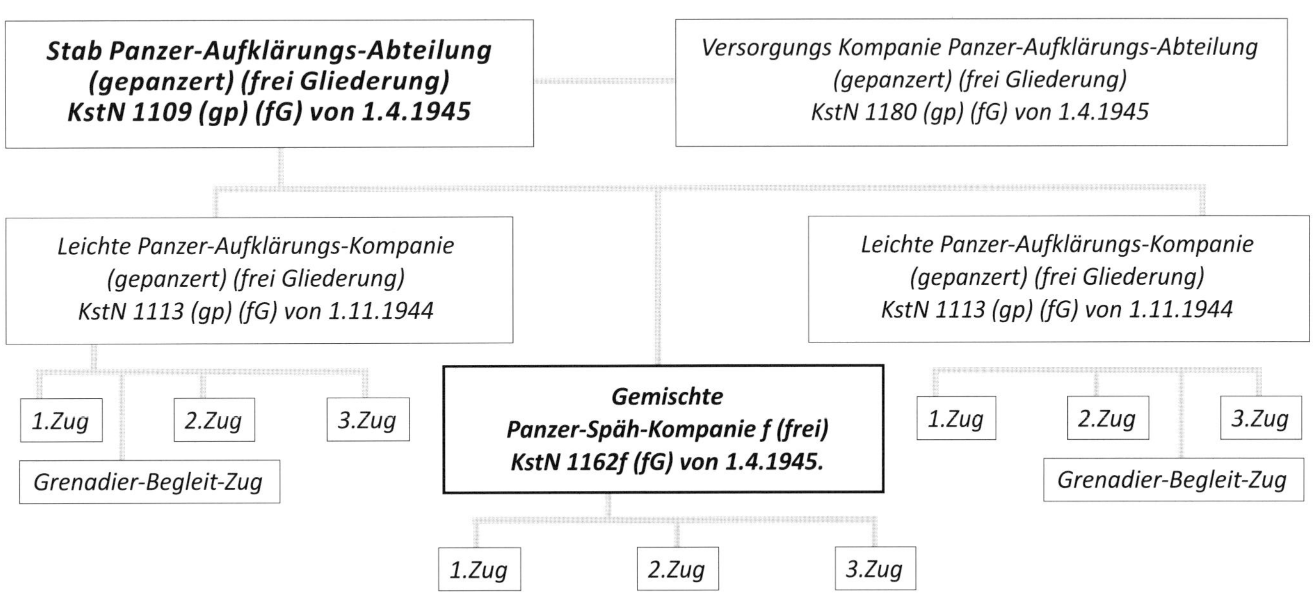

Note that the Kampfgruppe Panzer Division 45 organisation allowed for only one Panzer-Aufklärungs company organised with KstN 1113 (gp) (fG) and not the two shown here for the full division.

As mentioned in the introduction to this book, the development of the Sdkfz 251/9 Kanonenwagen was an attempt to provide the Panzergrenadiers, or armoured infantry, and reconnaissance units of the Panzer divisions with increased mobile firepower. After the relatively easy victories over Poland and France the German Armed Forces found themselves enmeshed in a conflict typified by staggering losses and vast distances. The initial lightning thrusts into Soviet Byelorussia and Ukraine of June and July 1941 had been spectacularly successful but the Red Army had not only continued to offer stubborn resistance but was able to field effective weapons of which the Wehrmacht had been largely unaware.

Within a short space of time, the role of the assault artillery and the heavier armoured units had shifted from infantry support to tank destruction, eventually resulting in the installation of larger and more powerful weapons. This left the Heereswaffenamt with surplus numbers of 7.5cm L/24 guns and it was soon

Above: Vehicle number WH-1286748, one of the two Büssing-NAG prototype Sdkfz 251/9 ausf C halftracks. Note that the radio antenna is fitted to the right side of the hull. Below: A 7.5cm-armed halftrack based on the Sdkfz 251 ausf D. This image provides a good view of the rectangular metal cradle (A) and gun shield (B). Note also the mount for the antenna insulator (C) on the hull side.

An Sdkfz 251/9 Kanonenwagen assembled prior to December 1944. Note the two-piece engine access doors. The shape of the aperture created for the gun cradle is well illustrated here. In the original print what appears to be a four-digit company number is just visible on the hull side.

Below: Two views of the multi-purpose 7.5cm KwK 51 (sf) gun mount. The complexity of the design is obvious in the overhead view. These mounts were also fitted to the Sdkfz 250/8 halftrack, as shown at right, and the Sdkfz 234/3 armoured car with a slightly modified shield.

soon decided that these could be fitted to suitably modified Sdkfz 233 armoured cars and Sdkfz 251 halftracks. These armoured halftracks had been in service since the Polish Campaign of late 1939 and were manufactured by Borgward and Hannoversche Maschinenbau AG, or Hanomag. The Berlin firm of Büssing-NAG was contracted to convert existing halftracks by fitting them with the 7.5cm gun and two prototype vehicles were sent to the Eastern Front in June 1942. The trials were successful and an initial order was placed for 150 vehicles.

By this time the Sdkfz 251 ausf C, with its simplified hull, had been in production for some time and this was the model chosen for the Kanonenwagen conversions. The manufacture of this variant continued until September 1943 and although sources differ, somewhere between 3,700 and 4,100 were built (1).

Although Hanomag remained the principal manufacturer of the Sdkfz 251 series throughout the war, the increased demand meant other firms were contracted to produce components and assembled vehicles. For example, complete hulls were manufactured by Ferrum AG, Böhmische Leipa, Steinmüller GmbH and Schöller Bleckmann AG in Austria. It was the latter firm which received one of the first orders for the Sdkfz 251/9 Kanonenwagen when it was contracted to build sixty complete vehicles in October 1942.

The conversion of the basic Sdkfz 251 halftrack involved removing the co-driver's visor and a sizeable section from the hull front and roof on the right side. This created an aperture for the barrel of the 7.5cm L/24 gun which was held in a rectangular metal cradle. This cradle was the major difference between the halftrack installation and that of the Pzkpfw IV and Sturmgeschütz III. An armoured shield sat above and on either side of the cradle and this was bolted to the hull on the inner edge. The main gun was fitted to a pedestal which was bolted to the hull floor, located behind and slightly to the right of the driver. The co-driver's seat was removed to accommodate the gun and the benches fitted to the standard vehicle were replaced by a folding seat on the right, above which the gun cleaning rods were stowed, and an ammunition locker on the left. The latter was situated directly behind the gunner (2).

On very early conversions, including the Büssing-NAG prototypes, a radio antenna and insulator were fitted to the right side of the hull near the gun breech. On later conversions the antenna fitting was moved to the left side. The 7.5cm gun could be

Notes

1. Most of the official documentation regarding these vehicles was lost or destroyed late in the war and much of the information presented here, particularly in reference to production, is based on the post-war accounts of Albert Speer. Fortunately we do have accurate figures for the numbers of vehicles fitted with the 7.5cm gun and these are presented on page 2.

2. The radio was moved from its normal position in front of the co-driver and situated between the gunner and the ammunition locker.

A view through the rear access doors of an Sdkfz 251/22 ausf D. Visible here are the main gun breech (A), the elevating and traversing gears (B), the driver's position (C) and instrument panel (D), ammunition lockers (E), individual ammunition canisters (F) balanced on the gunner's seat and the bench (G). Note how the lower edges of the gun shield have been cut to fit the hull.

traversed 10° to the left and 12° to the right. As with the Sturmgeschütz III assault gun the driver did his best to align the vehicle with the target. The aiming device, with its Sfl.Z.F. 1 sight, and the traversing and elevating gears were all situated on the left side of the gun as was the gunner's seat.

The practical range of the 7.5cm L/24 gun was up to 6,000 metres for the high explosive and smoke rounds while the Panzergranate 38 armour-piercing shell was capable of destroying a lightly armoured vehicle at 800 metres. In reality the ranges at which targets were engaged were much less and it was never intended that the Kanonenwagen act as an anti-tank vehicle. In early 1943 the Sdkfz 251 ausf D halftrack went into production and from July this all-welded version replaced the earlier model as the basis for the Kanonenwagen conversions. The installation of the 7.5cm gun and the internal modifications were unchanged and over 200 vehicles could have been built in this configuration.

In December a multi-purpose gun mount, referred to as the 7.5cm KwK 51(Sf), was introduced. This featured higher armoured sides which offered increased protection to the crew but did raise the vehicle's profile considerably. The gun was fitted with a coaxial MG 42 and was fixed directly onto a reinforced roof on the hull front. The armoured sides were bolted to the hull and although their construction was somewhat complicated, the new arrangement meant that it was no longer necessary to cut into the armour plate of the halftrack.

Importantly, the gun and its mount could be fitted by units in the field and it was also issued in kit form. Interestingly, supplies of the 7.5cm L/24 gun had all but run out by this time but the Kanonenwagen conversions had been so successful that its production was resumed.

By late 1944, anticipating that production of the Pzkpfw IV would cease and therefore alleviate the need for the tank's 7.5cm L/48 gun, Hitler ordered that priority should be given to the manufacture of the towed 7.5cm L/46 PaK 40 anti-tank gun and that these weapons, suitably converted, be fitted to as many vehicles as possible (1).

In less than two months the first production models had left the factory and a number of conversions kits, which were intended to replace the weapons fitted to Sdkfz 251/9 halftracks that were already in service, had been sent to operational units. All factory-built conversions were based on the Sdkfz 251 ausf D model. But development of the Sdkfz 251/22 was rushed and the weight of the main gun would cause technical problems which were never overcome. The driver's position remained unchanged from the Sdkfz 251/9 and, as with the first conversions, there was no longer a need for the co-driver and his visor was replaced by a metal plate. Photographs of Sdkfz 251/22 halftracks fitted with the standard visor in both positions are probably conversions built in the field. A gun travel lock was fitted to the hull front between the visors and a flattened U-shaped section was cut from the roof to accommodate the gun's recuperator. The size of the 7.5cm L/46 gun and its recoil necessitated the fitting a special frame, made up from steel beams and parts of the gun's original undercarriage, which was installed behind the driver who was now forced to crawl under the gun to reach his seat. Two ammunitions lockers were built into the frame, one on the right fitting into the hull side and holding seventeen rounds and another under the gun which held five rounds.

Spare rounds were carried in individual canisters on the floor and a small folding seat was provided for the gunner on the left side. Behind this was a single rifle rack, which was used for any kind of stowage, and a small bench. Except for the PaK 40 carriage the gun was identical to the towed version and the left and right traverse was limited to 18° and 19° respectively.

Early production Sdkfz 251 Ausf D halftracks had position lights on the left and right front fenders, which were soon phased out in an effort to speed production, and the Notek driving light of the earlier variants which was replaced, probably from early 1944, by a Bosch headlight. Many late production vehicles had a prominent reflector fitted under the left rear track guard. Very late in the production of these halftracks a single-piece engine access door was introduced and the late pattern steel tracks can be seen on a number of Sdkfz 251/22 vehicles photographed in 1945.

Notes

1. The main physical difference in these weapons lay in the length of the breech. The ammunition fired by the PaK 40 had a larger charge and therefore slightly more penetrative power. They were also available in large numbers and during 1944 complete guns were being assembled at an average of 1,000 per month.

An Sdkfz 251/22 ausf D photographed just after the end of the war. Note the single-piece engine access door with hinges along the top edge (A), which is actually missing here, the position of the antenna insulator (B), the metal plate covering the co-driver's visor (C) and the gun travel lock (D).

Dragon Models Ltd
B1-10/F., 603-609 Castle Peak Rd,
Kong Nam Industrial Building,
Tsuen Wan, N. T., Hong Kong
www.dragon-models.com

Tamiya Inc
Shizuoka City, Japan
www.tamiya.com

Italeri S.p.A.
via Pradazzo 6/b,
40012 Calderara di Reno, Bologna, Italy
www.italeri.com

Revell
Revell USA
www.revell.com
Revell Germany
www.revell.de

AFV Club/Hobby Fan
6f., No.183, Sec.1, Datong Rd, Xizhi City,
Taipei County 221, Taiwan
www.hobbyfan.com

Armourfast
Unit 4, Orsett Industrial Park,
Stanford Road, Orsett, Grays
RM 16 3BX, United Kingdom
www.armourfast.com

Voyager
Room 501, No.411 4th Village,
SPC Jinshan District, Shanghai 200540,
P.R.China
www.voyagermodel.com

Aber
ul. Jalowcowa 15, 40-750 Katowice,
Poland
www.aber.net.pl

Although the companies listed below are not mentioned elsewhere in this book they all produce tracks sets for the Sdkfz 251 series.

Friulmodel
H 8142. Urhida, Nefelejcs u. 2., Hungary
www.friulmodel.hu

Modelkasten
Chiyoda-ku Kanda, Nishiki-Cho 1-7, Tokyo,
Japan
www.modelkasten.com
Very difficult to navigate but worthwhile.

Master Club
www.masterclub.ru
It appears that this firm is closely associated with Armour35, a Russsian mail-order firm.

Eduard Model Accessories
Mirova 170, 435 21 Obrnice,
Czech Republic
www.eduard.com

Al-By
17, rue des Limousins,
82400 Valence d'Agen,
France
www.albyminiatures.e-monsite.com

Gaso.Line
1/48 scale models and accessories
www.gaso-line.eu

Royal Model
Via E. Montale, 19-95030 Pedara, Italy
www.royalmodel.com

RB Model
Powstancow Wlkp.29B,
64-360 Zbaszyn,
Poland
www.rbmodel.com

Hasegawa
3-1-2 Yagusa, Yaizu 425-8711,
Shizuoka City, Japan
www.hasegawa-model.co.jp

As with the books in the TankCraft series I have endeavoured to give the reader an understanding of the development and employment of the infantry support vehicles and tank destroyers based on the Sdkfz 251 halftracks while also providing the modeller with the inspiration to create a scale representation. In researching this project I consulted a number of works including the books by Detlev Terlisten in the Nuts & Bolts series, *Sdkfz 251 Ausf D* by Hans-Heiri Stapfer and *Panzer Divisions 1944-1945* by Pier Paolo Battistelli. I would very much like to acknowledge the contributors to the Axis History and the Feldgrau forums, especially the late Ron Klages and Martin Block whose research on unit histories and vehicle allocations would fill many volumes. I would also like to thank the modellers who very generously allowed me to publish the images of their work and my friend Ivano Biazzo who helped with the research for the colour illustrations. As always I am indebted to my friends and long-time associates Karl Berne, Valeri Polokov and J.Howard Parker for their invaluable assistance with the photographs and period insignia.

The view from the fighting compartment of an Sdkfz 251/9 ausf D Kanonenwagen of SS-Panzergrenadier-Regiment 9 Germania. Note the method of fixing the gun shield to the hull and the sight of the 7.5cm L/24 gun.